ODE ARCHITECT
Companion

ODE ARCHITECT
Companion

C·ODE·E
(Consortium for ODE Experiments)

JOHN WILEY & SONS, INC.

New York / Chichester / Weinheim / Brisbane / Singapore / Toronto

ISBN 0-471-12132-0

Printed in the United States of America

10 9 8 7 6 5 4 3 2 1

Printed and bound by Bradford & Bigelow, Inc.

PREFACE

This workbook was designed to accompany the software package ODE Architect, and that's why we call it a *Companion*. Each of the 13 Companion chapters corresponds to a multimedia module in the Architect and provides background and opportunities for you to extend the ideas contained in the module. Each chapter ends with several problem sets, called Explorations, related to the chapter and module topics. The Explorations are designed so that you can write in answers and derivations, and, since they are printed on perforated pages, they can be removed and handed in along with printouts of graphs produced by the Architect. There is also a notepad facility in the Architect which, with the cut and paste features, makes it possible to write reports.

ODE Architect

ODE Architect provides a highly interactive environment for constructing and exploring your own mathematical models of real-world phenomena, whether they lead to linear or nonlinear systems of ODEs. The Architect's multimedia front end guides you through experiments to build and explore your own ODEs. The software has numerical solvers, 2- and 3-D graphics, and the ability to build physical representations of systems such as pendulums and spring-mass systems as well as the ability to animate them. Together with its library of ODEs, the ODE Architect brings a wealth of opportunities to gain insights about solutions to ODEs.

The overall guiding feature is for the software to be easy to use. Navigational paths are clearly marked and simple to follow. When starting the software, you are presented with a title screen followed by a main menu allowing selection of a specific module. You may prefer to go directly to the Architect Tool to run your own experiments. At any place in the software, you will be able to call up the contents menus and access the material in any order. We expect that most will work through the multimedia modules. Let's look at each of the three principal parts of the ODE Architect in more detail: The Multimedia ODE Architect, The ODE Architect Tool, and the ODE Library.

Multimedia ODE Architect

C·ODE·E members and colleagues have authored the multimedia modules, each with its own theme. The modeling process is detailed, supported by highly interactive simulations. Students explore the problem-solving process via "what-if" scenarios and exercises. They are guided to build their own ODEs and solve them numerically and graphically, and compare the predicted results to empirical data when appropriate.

Each module has up to four submodules, and they range from the straight-forward to the advanced. The animations are often funny, the voice-overs and text informal, but the modeling and the mathematics are the real thing. Most submodules go through a model-building process and several experiment screens, and then end with some questions (Things-to-Think-About, or TTAs). These questions extend the topics of the submodule and take you to the solver tool to produce solution curves and orbits, or write a report connecting the mathematics, the models, and the pictures. When you open the Tool using a TTA link, the pertinent equations and parameter settings will automatically be entered into the equation quadrant of the Tool. You are then poised to think about, and without constraint, explore the model introduced in the submodule.

ODE Architect Tool

The ODE Architect Tool is a first-rate, research-quality numerical ODE solver and graphics package. The ODE Architect Tool employs a graphical user interface to enter and edit equations, control solver settings and features, and to create and edit a wide variety of graphics. A second mode of operation, the Expert Mode, provides access to more advanced features.

The Tool is the heart of the software, and it is a workspace where you can:

- Construct, solve, and explore ODEs
- Input data tables
- Graph and animate solution curves, phase plane graphs, 3D graphs, Poincaré sections, discrete maps, direction fields, etc.
- Build, analyze, and animate physical representations of dynamical systems.

The robust Tool will solve systems of up to 10 first-order ODEs which can be entered using a simple, natural scripting language. Auxiliary functions involving the state variables can be defined. A solver/grapher feature for discrete dynamical systems is also available from the Tool. A variety of engineering functions such as square waves, sawtooth waves, and step functions are included in the Tool function library. Two- and three-dimensional graphics are supported, as well as time and parameter animations of solution data. Initial conditions can be entered by clicking in a graph window or via the keyboard. Graph scales can be set automatically or manually. Numerical values of solutions can be viewed in tabular form. Parameter-sensitive analysis is made easy with a built-in parameter-sweep tool. You can do parameter and initial-value sweeps to see the effects of data changes on orbits and solution curves. Graphs are editable and you can scale and label axes, mark equidistant-in-time orbital points, color the graphs, change line styles, overlay graphs of functions and solution curves for different ODEs—all with no programming or special commands to remember.

The solvers in the ODE Architect are state-of-the-art numerical solvers based on those developed by Dr. L.F. Shampine and Dr. I. Gladwell at Southern Methodist University. For a delightfully readable account on using numerical ODE solvers in teaching ODEs, please refer to their paper:

> Shampine, L.F., and Gladwell, I., "Teaching Numerical Methods in ODE Courses"

in the book *Differential Equations for the Next Millennium*, edited by Michael J. Kallaher in the MAA Notes series.

Module 1, "Modeling with the ODE Architect", is an on-line tutorial for many of the features of the Tool. The Architect also has help facilities and the multimedia side is self-documenting.

ODE Library

The ODE Library has dozens of pre-programmed, editable, and interactive ODE files covering a wide range of topics from mathematics, physics, chemistry, population biology, and epidemiology. There are also many ODEs to illustrate points such as data compression, ODEs with singular coefficients, bifurcations, limit cycles, and so on. Each Library file has explanatory text along with the equations and includes an illustrative graph or graphs. The Library files are organized into folders by topic and they have descriptive titles to facilitate browsing. These files also provide a marvelous way to learn how to use the tool.

ACKNOWLEDGMENTS

ODE Architect was developed with partial support from the NSF/DUE,[1] by the Consortium for ODE Experiments (C·ODE·E), Intellipro, Inc., and John Wiley & Sons, Inc. C·ODE·E saw to the mathematical side of things, Intellipro rendered C·ODE·E's work into an attractive multimedia software package, and John Wiley coordinated the efforts of both teams.

As in any project like this, we owe a debt of gratitude to many people: reviewers, beta testers, students, programmers, and designers. Specifically, we want to thank the other members of the C·ODE·E Evaluation Committee, Barbara Holland (John Wiley & Sons), Philippe Marchal (Intellipro), Michael Moody (Harvey Mudd College), and Beverly West (Cornell University). Without the many hours of hard work they put in on this project, it could not have been done.

We especially want to thank Professor L.F. Shampine for providing the excellent solver codes (developed by himself and his colleague, Ian Gladwell), and for his continuing support of the project. Thanks to Mark DeMichele at Intellipro, who wrote the code for the Architect and the implementation of the Shampine/Gladstone solver codes, and who put up with our constant "advice". Another very special thanks to David Richards for designing and implementing the LaTeX macros for the Companion book and for his patience and meticulous attention to detail during the many revisions. We also very much appreciate the reviewers, editors, and evaluators Susan Gerstein, Zaven Karian, Mario Martelli, Lang Moore, Douglas Quinney, David Cook, and Robert Styer for their many helpful comments and suggestions.

Finally, a big "thank you" to our students Tiffany Arnal, Claire Launay, Nathan Jakubiak, Joel Miller, Justin Radick, Treasa Sweek, and many others who read chapters, tested modules, and commented freely (even favorably) on what they experienced.

Robert L. Borrelli
Courtney S. Coleman
Claremont, CA

[1]The work on ODE Architect and its Companion book was supported in part by the National Science Foundation under Grant Numbers DMI-9509135 and DUE-9450742. Any opinions, findings, conclusions, or recommendations expressed in this material are those of the authors and do not necessarily reflect the views of the National Science Foundation.

INFORMATION ABOUT MODULES/CHAPTERS

Overview

Modules/Chapters 1–3 are all introductory modules for first-order ODEs and simple systems of ODEs. Any of these modules/chapters can be used at the beginning of an ODE course, or at appropriate places in elementary calculus courses.

Modules/Chapters 4–9 involve higher-order ODEs and systems and their applications. Once students understand how to deal with two-dimensional systems graphically, any of these modules/chapters is easily accessible.

Modules/Chapters 10–12 apply two-dimensional systems to models that illustrate more advanced techniques and theory; the multimedia approach makes them nevertheless quite accessible. The modules are intended to enable students to get much further with the technical aspects explained in the chapters than would be otherwise possible.

Module/Chapter 13 treats discrete dynamical systems in an introductory fashion that could be used in a course in ODEs, calculus, or even a non-calculus course.

A Multimedia appendix on numerical methods gives insight into the ways in which numerical solutions are constructed.

Description/Prerequisites for Individual Modules/Chapters

We list below for each Module/Chapter its prerequisites and some comments on its level and goals. In general, each module progresses from easier to harder submodules, but the first section of nearly every module is at an introductory level.

The modules can be accessed in different orders. It is not expected that they will be assigned in numerical order. Consequently, we have tried to explain each concept wherever it appears, or to indicate where an explanation is provided. For example, Newton's second law, $F = ma$, is described every time it is invoked.

There is far more material in ODE Architect than could possibly fit into a single course.

Module/Chapter 1: Modeling with the ODE Architect

Assumed concepts: Precalculus; derivative as a rate of change

This module is unlike all the others in that it is not divided into submodules, and it provides a tutorial for learning how to navigate ODE Architect. It carries that tutorial process along in tandem with an introduction to modeling that assumes very little background.

Module/Chapter 2: Introduction to ODEs

Assumed concepts: Derivatives; slopes; slope fields

The module begins with some simple first-order ODEs and their solutions and continues with slope fields (and a slope field game).

The Juggler and the Sky Diver submodules use second-order differential equations, but both the chapter and the module explain the transformation to systems of two first-order differential equations.

Module/Chapter 3: Some Cool ODEs

Assumed concepts: Basic concepts of first-order ODEs, solutions, and solution curves

Newton's law of cooling, and solving the resulting ODEs by separation of variables or as linear equations with integrating factors, are presented thoroughly enough that there need be no prerequisites.

The submodule for Cooling a House extends Newton's law of cooling to real world cases that are easily handled by ODE Architect (and not so easily by traditional methods). This section makes the point that rate equations and numerical solutions are often a much smarter way to go than to trudge toward a solution formula.

Module/Chapter 4: Second-Order Linear Equations

Assumed concepts: Euler's formula for complex exponentials

The module and chapter treat only constant coefficient ODEs. The chapter begins by demonstrating how to treat a second-order ODE as a system of first-order ODEs which can be entered in ODE Architect. Both the first submodule and the chapter explain from scratch all the traditional details of an oscillating system such as amplitude, period, frequency, damping, forcing, and beats.

The Seismograph submodule is a real world application. The derivation of the equation of motion is not simple, but the multimedia module gives insight into the workings of a seismograph, and it is not necessary to understand the details of the derivation to use and explore the modeling ODE.

Module/Chapter 5: Models of Motion

Assumed concepts: Newton's second law of motion

This module's collection of models of motion in one and two dimensions is supported by a chapter that gives background on vectors, forces, Newton's laws, and the details of the specific submodules; so it stands on its own without further prerequisites.

Module/Chapter 6: First-Order Linear Systems

Assumed concepts: Basic matrix notation and operations (multiplication, determinants); complex numbers; Euler's formula

This unit introduces all of the basic notions, both algebraic (emphasized in the chapter) and geometric (emphasized in the module), for linear systems. The central roles of eigenvalues and eigenvectors are explained. The Tool can be used to calculate eigenvalues and eigenvectors.

The Explorations bring in coupled tank problems (Chapter 8 introduces compartment models) and small motions of a double pendulum (which are extended in Chapter 7).

Module/Chapter 7: Nonlinear Systems

Assumed concepts: Equilibrium points; phase plane and component plots; matrices; eigenvalues and eigenvectors

The goal is to use graphical solutions to make handling nonlinear systems as easy (almost) as linear systems. Linearization of a nonlinear ODE is introduced as a basic concept, and the chapter goes on to elaborate perturbations and bifurcations. The Tool can be used to find equilibrium points, and calculate the Jacobian matrix and its eigenvalues/eigenvectors at each equilibrium point. The predator-prey and saxophone reed models are introduced and explained in the module while the spinning bodies and double pendulum models are treated in the chapter and also in the Library with an animated model linked to the ODE.

Module/Chapter 8: Compartment Models

Assumed concepts: Systems of ODEs

Both the module and the chapter use 1D, 2D, 3D, and 4D applications (in sequence) to illustrate principles of the Balance Law and interpretations of solutions. The final submodule introduces Hopf bifurcations and the interesting behavior of chemical reactions in an autocatalator. Three of the models are linear; the last is nonlinear.

Module/Chapter 9: Population Models

Assumed concepts: Systems of ODEs

The module and chapter introduce simple 1D, 2D, and 3D nonlinear models, and give a discussion of the biology behind the models.

Module/Chapter 10: The Pendulum and Its Friends

Assumed concepts: Systems of ODEs; the first submodule of Module 4; the arctangent function; parametric curves on a surface

The pendulum submodule explores all the traditional aspects of a pendulum, using integrals of motion. Child on a Swing and Geodesics on a Torus give new extensions of pendulum analysis; supporting detail is given in the chapter. The approach to modeling is a little different in this chapter—for example, how to invent functions that behave as needed (Child on a Swing), or how to exploit part of an ODE that looks familiar (Geodesics on a Torus).

Module/Chapter 11: Applications of Series Solutions

Assumed concepts: Systems of ODEs; acquaintance with infinite series and convergence; the first submodule of Module 4

The module introduces the techniques and limitations of series solutions of second-order linear ODEs. The Robot and Egg provides motivation for the subject and Aging Springs illustrates Bessel functions. The chapter contains information about the mathematics of series solutions.

Module/Chapter 12: Chaos and Control

Assumed concepts: The pendulum ODEs of Module 10; systems of ODEs; experience with Poincaré sections and/or discrete dynamical systems (Chapter 13) is helpful

The three submodules of this unit tell a story, and in the process illustrate a theorem from current research. This module uses sensitivity to initial conditions and the Poincaré section to assist with the analysis. Sinks, saddles, basins, and stability are described. Finally, the elusive boundaries of the Tangled Basin provide a mechanism for control of the chaotically wandering pendulum. The module ends in a fascinating control game that is both fun to play and illuminates the theorem mentioned above.

Module/Chapter 13: Discrete Dynamical Systems

Assumed concepts: Acquaintance with complex numbers and the ideas of equilibrium and stability are helpful

The module provides a gentle introduction to an increasingly important subject. The chapter fills in the technical and mathematical background.

This module could be used successfully in a liberal arts course for students with no calculus.

Level-of-Difficulty of Modules

The chart below is a handy reference for the levels of the submodules.

Elementary		Intermediate		Advanced
1				
2.1, 2.2	2.3, 2.4			
3.1	3.3	3.2		
4.1				4.2
5.1	5.2	5.3		
	6.1	6.2	6.3	
	7.1		7.2	7.3
8.1	8.2	8.3	8.4	
9.1	9.2	9.3		
		10.1	10.2	10.3
		11.1	11.2	
		12.1		12.2, 12.3
13.1	13.2		13.3	

In constructing this chart we have used the following criteria:

Elementary: Straightforward, self-contained, can be used as a unit in any introductory calculus or ODE course.

Intermediate: Builds on some prior experience, including earlier submodules and chapters.

Advanced: More challenging models or mathematics, especially suitable for term or group projects.

CONTENTS

8   Compartment Models   135

Courtney Coleman & Michael Moody, Harvey Mudd College

9   Population Models   155

Michael Moody, Harvey Mudd College

10   The Pendulum and Its Friends   173

John Hubbard & Beverly West, Cornell University

1

Modeling with the ODE Architect

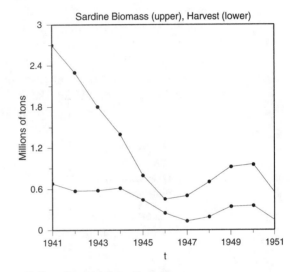

Pacific sardine population and harvest.

Overview   In the two decades from 1932 to 1951, the Pacific sardine fishery completely collapsed. In this chapter you will learn to use the ODE Architect to construct a mathematical model which describes this event rather well. This will have two purposes: it will familiarize you with the menus and features of the ODE Architect, and it will acquaint you with the principles of mathematical modeling.

    First we'll construct a model for the Pacific sardine population during the years 1930–1950 as if it were unharvested. Then we will focus on the harvesting that actually took place and see how it contributed greatly to the destruction of the sardine population.

Key words   Modeling; Pacific sardine; population model; initial conditions; exponential growth; carrying capacity; logistic equation; harvesting

See also   Chapter 9 for more on population models.

◆ Building a Model of the Pacific Sardine Population

Step 1: State the problem and its context

☞ In this chapter we will build a model for the sardine fishery in California and also introduce features of the ODE Architect Tool for solving differential equations. Consult the *User's Guide* for a full description of all features of the tool.

The Pacific sardine (*Sardinops sagax caerulea*) has historically experienced long-range cycles of abundance and depletion off the West Coast of California. It was during one of the abundant periods, 1920 through 1951, that a huge sardine fishing and canning industry developed. The total catch for the California coastline reached a peak of 726,124 tons during the 1936–37 season (June through the following May). The Pacific sardine population then began a serious decline during the 1940s until, as one estimate has it, by 1959 the sardine biomass was 5% (0.2 million tons) of the 1934 level (4 million tons). (The *biomass* is the amount of a particular organism in its habitat.) There is general agreement that heavy harvesting played a role in the decimation of the Pacific sardine during that period. The fishing industry had a serious decline after the 1950–51 season: increasing numbers of fishermen went bankrupt or moved to other fisheries. Undoubtedly the canneries were also affected.

After 50 years of fishing for the Pacific sardine, a moratorium was imposed by the California legislature in 1967. The Pacific sardine seems to be making a comeback as of the mid-1980s, though the numbers are not yet near the abundant levels of the 1930s.

Here are the goals of your model:

☞ Problem statement.

1. Determine the extent to which the precipitous decline of the Pacific sardine population was due to over-harvesting from 1941 to 1951.

2. Ascertain an optimal harvest rate that would stabilize and sustain the sardine population during that time period.

Step 2: Identify and assign variables

Assigning the variables in a mathematical model is a a skill that requires some practice. Doing some background reading and studying the context of the problem and the problem statement helps to clarify which are the most important features of the system you wish to model.

It turns out that there tend to be long-range cycles of Pacific sardine abundance and scarcity. These cycles are not yet completely understood, but it is certain that factors such as ocean temperature, nutrient upwelling from deep waters, currents that aid fish migration, predator populations of larger fish and sea lions, and, of course, fishing, play a vital role in the cycles. In this model we will focus on a period when the harvesting of the sardine was very heavy. Due to the large magnitude of the harvesting, its effect was dominant for the period of time we will model, 1941–1951, so we will neglect the other factors. That the other factors still operate on the population is evidenced by the difficulty in getting the model to match the data perfectly. Nevertheless, you'll see how modeling, while not always explaining every aspect, provides insight into the dynamics of an otherwise very complex biological relationship.

Given the information we have at this point we need the following variables and parameters in our model:

☞ Sardine biomass is a state variable, the other quantities are parameters.

1. Sardine biomass (in units of million tons)
2. Growth rate for the Pacific sardine (in units of million tons/year)
3. Maximum biomass, or *carrying capacity* (in units of million tons)
4. Sardine harvesting (in units of million tons/year)

Note that we opted to define sardine biomass in million tons, rather than numbers of fish, to be consistent with the data and estimates used.

It's good practice to introduce as few parameters as necessary into a model at first. Additional parameters can be added if they are needed to improve the accuracy of the model. The model may be refined until the desired level of accuracy is achieved.

With the variables and parameters identified, the next step is to construct an equation for the rate of change of the state variable in terms of the state variable itself, the model's parameters, and possibly also time. This equation is known as a differential equation (abbreviated ODE). When an ODE is entered into ODE Architect along with an initial value of the state variable, the Architect Tool displays a graphical representation of the solution.

☞ Graphical representation of the problem.

Figure 1.1 represents the estimated Pacific sardine biomass and harvest during the period 1941–1951. Note again the use of sardine biomass in million tons, rather than numbers of fish.

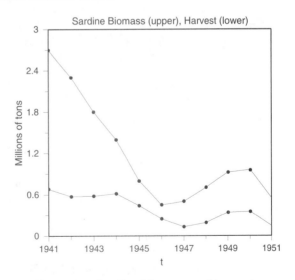

Figure 1.1: Sardine biomass and harvest.

Our first task is to use ODE Architect to build a mathematical model to simulate the growth and decline of the Pacific sardine biomass without harvesting. Then we can explore the impact of harvesting on that biomass.

☞ Click on the spinning orb to go directly to the Architect Tool.

Place the ODE Architect CD-ROM in your computer and start the ODE Architect Tool. Four quadrants will be displayed on the screen (see Figure 1.2). The upper left quadrant is the equation quadrant; it should be empty now. The two right-hand quadrants should be empty. These are plot quadrants that will display 2- or 3-dimensional plots when you solve differential equations. The lower left quadrant currently shows the initial conditions (**IC**) display. Notice that it is selectable using the four tabs (**IC**, **Sweep**, **Solver**, **Equilibrium**) on the lower edge of the quadrant. For now leave **IC** selected.

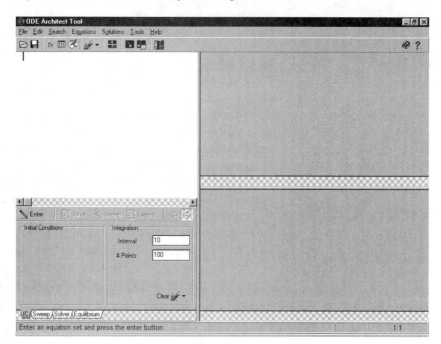

Figure 1.2: ODE Architect tool screen

Step 3: State the relationships that govern the variables
We begin by simulating the unchecked growth (no harvesting) of the Pacific sardine population, which we will designate as the state variable *sardine*. Basic biology suggests that it is reasonable to assume that the rate of biomass growth (i.e., the derivative $d(sardine)/dt$) at a given time t is *proportional* to the quantity of sardines (the size of the biomass) present at that time t.

Step 4: Translate the laws into equations
Since *sardine'* is a common notation for the derivative (rate of change) of *sardine*, we can write

$$sardine' = r * sardine$$

where r is a proportionality factor that we will refer to as the *growth rate factor*.

Fishery and biomass data collected over the period 1932–1958 indicate that the Pacific sardine population has had a volatile history. The Pacific sardine biomass, if not manipulated or constrained, can grow at a rate of between 10% and 40% per year. We will assume a moderate position and set the growth rate factor at $r = 0.20$. A modeler often has to make assumptions and guess parameter values to get a model started; you can refine the assumptions later.

Step 5: Solve the resulting differential equations

☞ Entering differential equations.

Point and click the cursor in the equation quadrant and type in

$$sardine' = r * sardine \qquad (1)$$

using an apostrophe for the prime, and an asterisk for multiplication; hit **Return** (or **Enter**) and assign the value 0.20 to the parameter r by typing in

$$r = 0.20$$

Now click the cursor on the box marked **Enter** just below the equation quadrant. Notice that this causes scales to appear in the two plot quadrants.

☞ Setting initial conditions.

Now go to the lower left quadrant to set the initial conditions. Double click in the appropriate box to select a variable; then type in the new value. Set **t** (time) to start at 1930 and set **sardine** to be 1 (unit of million tons).

☞ Setting the time interval.

We'll go back later and put in a more realistic estimate for *sardine*. In the **Integration** panel, set the solve time to 20 by inserting the number 20 in the **Interval** box. Leave the default value of 100 in the **# Points** box.

☞ Starting the solution.

Click the **Solve** icon and notice that the right arrow is automatically selected. Your screen will look something like this (Figure 1.3):

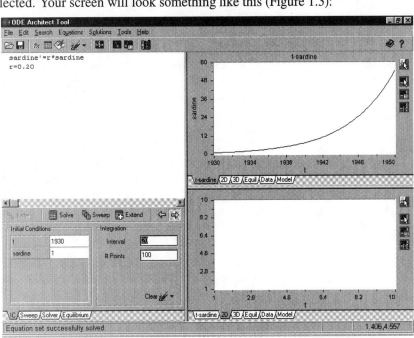

Figure 1.3: Exponentially growing sardine population.

☞ Unbounded growth.

Step 6: Interpret and test the solutions in context

There is now a classical exponential growth curve in the upper right quadrant. This implies that the sardine biomass grows without bound, which can't be true as there is not enough room on the planet! The exponential growth must be limited by factors like available food supplies, disease, predators, and so on; therefore we have to modify our model to reflect this fact. We learned earlier that the sardine biomass has been as large as 4 million tons, but we don't truly know the maximum sustainable biomass (*carrying capacity*), so to start let's assume a carrying capacity of 6 million tons. We can refine this guess later if we have trouble fitting the model to actual data. As we said before, it's not uncommon to have to make informed guesses for values that are not known or available. Then the values can perhaps be deduced by "fine tuning" (refining) the model in subsequent iterations to conform to reality.

Step 7: Refine the model to predict the empirical data

☞ Introduce a carrying capacity.

The following differential equation is sometimes used to exhibit maximum carrying capacity behavior in a population:

$$sardine' = r(capacity - sardine)$$

This equation says that the growth rate at any time is proportional to the "room to grow" factor $capacity - sardine$. Now click the cursor in the equations region. Using our assumed growth rate constant of $r = 0.20$ per year and a carrying capacity of 6 million tons, we modify the sardine growth to be

$$sardine' = r * (6 - sardine) \qquad (2)$$

Before entering the new sardine ODE, clear the graphics screens by clicking on **Clear** at the lower left and choosing **Clear All Runs**. A "confirmation" window will pop up; click on **Yes**. Now click in the equation region, make the corrections to your equation, and then click on the box marked **Enter**. Click on the **Solve** icon and notice in the plot window that the graph of the sardine biomass climbs and levels out at the assumed carrying capacity of 6 million tons.

✓ "Check" your understanding by comparing this curve with the earlier one and notice some significant differences: (1.) The first curve was concave up; this one is concave down. Why is that significant? (2.) The first curve grew without bound and had no asymptotes; the second curve has a horizontal asymptote. Explain why.

Examine the two graphs carefully at early values of t, say the first five years. Recall that the slope of a line tangent to the solution curve is the growth rate of the biomass at that time. How do the two curves differ in this regard? When is the rate of change of the biomass the greatest? Is it realistic for a biomass to exhibit its greatest rate of increase when the population is smallest? The answer to these questions is not as simple as you might think. For many biological populations, rate of change is proportional to the size of the population. The solution of ODE (1) exhibits this proportionality but it is unconstrained and so it's not useful over its whole domain. The solution curve of ODE (2) doesn't exhibit the proportional growth property. Which of the two is most appropriate for the Pacific sardine? We'll come back to that question after a little exploration with ODE (2).

☞ Exceed carrying capacity.

Now let's see if ODE (2) will allow us to exceed the carrying capacity for any length of time. Change the initial biomass to 12 million tons of sardines in the IC window. Click on the **Solve** icon. Notice that the vertical scale in the graph changes to accommodate the revised values and that both the old (lower) and the new (upper) curves are displayed on the graph. Observe what happened to the "overstocked" sardine population. How does this compare to what happened when the initial sardine biomass was 1 million tons? If you examine the two plots closely, you'll see that both plots stabilize at a level of about 6 million tons (see Figure 1.4).

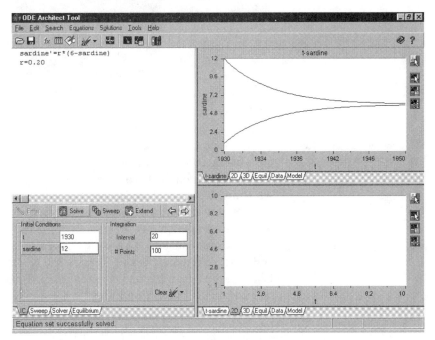

Figure 1.4: Sardine populations approach carrying capacity.

☞ To set scaling of axes.

Sometimes it's advantageous to change the scales of the axes to make graphs easier to read and interpolate, so we'd like to show you how to reset the vertical and horizontal scales. (The default setting for scales for ODE Architect is **Auto Scale**.) Select the upper right graph by placing the cursor arrow on the graph and clicking the *right-most* mouse button (or click on the icon at the upper right corner of the graph). You will see various plot window options presented. Select **Scales** from the resulting dropdown menu using the *left-most* mouse button. Click on **Auto Scale** to toggle it off. (The check in the box will disappear.) To change other values, double click in the box to select the value and just type to make a change. On the **X-Scale** menu set **Minimum** = 1930; **Maximum** = 1950; **Number of Ticks** = 10; and **Label every** = 2. (Adjust the number of ticks by clicking on the down arrow and selecting, or by double clicking the box and typing in the new value.) Make sure **Linear** is selected (not **Log**). Now select the **Y-Scale** menu (at the top): click the **Auto Scale** to toggle it off; set **Minimum** = 0; **Maximum** = 12; **Number of Ticks** = 10; and **Label every** = 2; and check that the **Linear** button is selected. Your screen should have a window that looks like Figure 1.5:

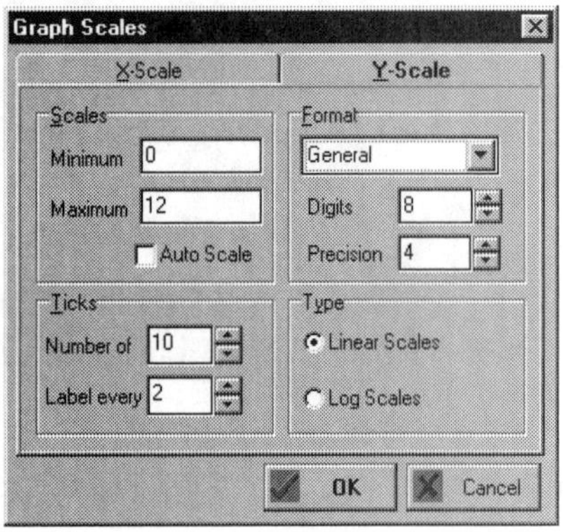

Figure 1.5: Plot scales window.

☞ You must click the **OK** button to enter changes.

Click on **OK** to cause the graph to be rescaled. (In this particular case, it turns out that the scale did not change from the automatically selected value.)

Now click on the **Clear** box in the **Integration** panel and choose **Clear All Runs**. What has changed? Next click on the **Solve** icon again. Notice that you got only the most recent curve (carrying capacity exceeded); you cleared the previous solution.

☞ Sweeping a variable.

We can extend the ease of making comparisons by sweeping through several possible initial values for *sardine* and displaying them all on one graph. Click on the **Clear** box and choose **Clear All Runs**. Now notice the **Sweep** tab beneath the **Initial Conditions** panel; click on **Sweep**. Click on **Single** for type of sweep. We will choose **Sweep 1** to be a sweep (multi-plot) of various initial values of *sardine*. In the **Sweep 1** box, click on the down arrow and select **sardine**. Set **Start** = 1; **Stop** = 7; and **# Points** = 3. Now click on the **Sweep** box next to the **Solve** icon (*not* the **Sweep** tab) (Figure 1.6).

☞ If the **# Points** is set to *n*, you'll get *n* overlayed graphs.

Notice that ODE Architect makes several runs. Notice also that the initial value for *sardine* located in the IC window was ignored and the values we entered in the sweep conditions were used instead.

To better see these results, let's rescale the vertical axis (**Y-Scale**) to **Minimum** = 1 and **Maximum** = 7. Look back at page 7 if you do not recall how to do this. Figure 1.7 shows that multiple runs are easily comparable in this format. Which initial value for *sardine* created the most stable or flattest curve? Does the population always stabilize around the same biomass?

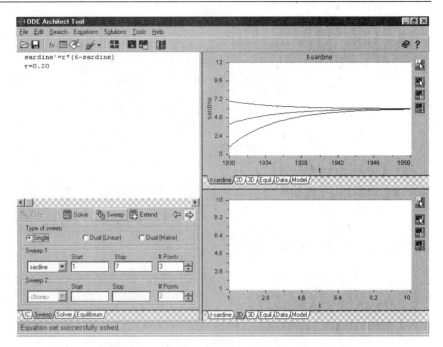

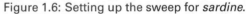

Figure 1.6: Setting up the sweep for *sardine*.

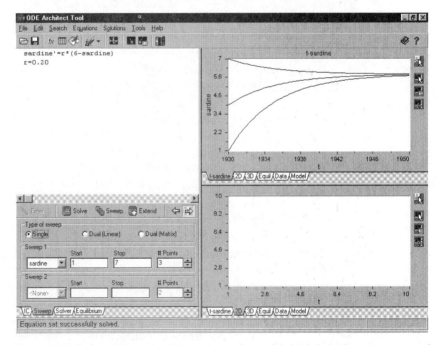

Figure 1.7: Sweeping and solving gives plots with initial sardine tonnages of 1, 4, 7 (in millions).

Do the resulting curves accurately represent the growth you'd expect over the whole range of values for *t*? Growth is usually proportional to population size when well below the carrying capacity. However, when you look at your graph notice that for small populations of sardines, the growth rate is rather steep. As the sardine population approaches the carrying capacity the biomass should level off, which the preceding curves do reflect.

Now we'll examine the properties of the model we created in ODE (2). The growth rate is proportional to (*capacity − sardine*) and so for small sardine biomass, the biomass grows at a nearly constant rate. Near the carrying capacity, the factor (*capacity − sardine*) causes a leveling off (see Figure 1.7): the factor forces growth to be proportional to the distance from capacity.

◆ The Logistic Equation

Combining the elements of the proportional growth model given by ODE (1) and the restricted growth model given by ODE (2) leads to what is called the *logistic equation for growth* (or the *Verhulst equation*, after the nineteenth century Belgian mathematician and biologist P. F. Verhulst):

$$sardine' = r \, sardine \frac{6 - sardine}{6} \tag{3}$$

Notice that for values of *sardine* very near zero, the factor $r * sardine$ dominates the computation, causing approximate exponential growth behavior. This is because the factor $(6 - sardine)/6$ has a value very near 1. For values of *sardine* near 6 (the carrying capacity), the factor $(6 - sardine)/6$ is near zero, and so growth slows to approach zero. Therefore we can expect exponential growth for small biomass with growth tapering off as the biomass approaches carrying capacity. Let's see if this refinement improves the model.

☞ Changing the equation.

Click on the **IC** tab to clear the graph and enter a new equation. After clicking on the **Clear** box, and choosing **Clear All Runs**, click in the equations quadrant and modify the growth ODE to read:

$$sardine' = r * sardine * (6 - sardine)/6 \tag{4}$$

Don't forget to click the box labeled **Enter**. Reset the initial sardine biomass to **1**. Finally, click on the **Solve** icon. Your screen should look something like Figure 1.8.

Notice that the graph now displays a mathematical representation more like what we expect of the sardine biomass over the long term. It is an elongated S-shaped curve with slow growth for small biomass, maximum growth near the midrange, and slow growth near the carrying capacity.

☞ Try various initial values. As before, use a **Single** sweep.

Use the sweep feature now to see how the logistic growth curve responds for various initial conditions for the variable *sardine*. **Sweep 1** *sardine*; **Start = 1**; **Stop = 7**; **# points = 4**. Click on **Sweep**. Figure 1.9 shows the four solution curves.

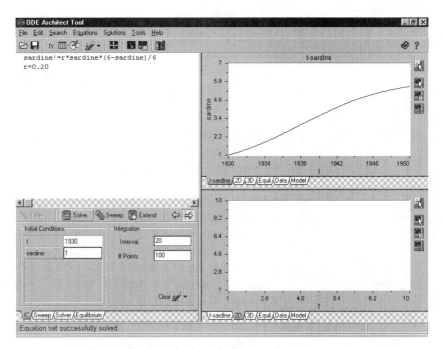

Figure 1.8: A logistic growth curve.

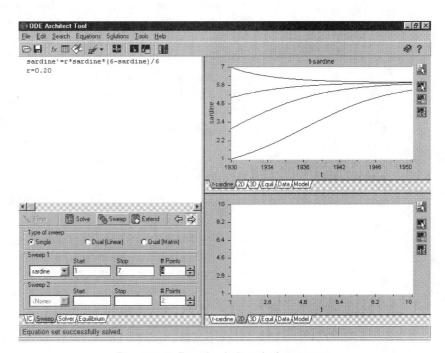

Figure 1.9: Four logistic solution curves.

✓   Does the model respond to your initial conditions in a reasonable manner? Do you think that this is a good population model to use for modeling the biomass of the Pacific sardine?

Let's now use the model given in equation (3) to explore the harvesting that took place in the years 1941–1951.

◆ Introducing Harvesting via Landing Data

In Figure 1.1 you saw a graph of the Pacific sardine harvest and the resulting biomass decline during the years 1941–51. We have not yet taken into account this harvest (or *landing*) data in our model; so our model does not yet reflect the collapse of the sardine fishery that occurred. We'll now incorporate the landing data into our model in the form of a lookup table.

The tutorial in Module 1 provides you with landing data for the Pacific sardine over the time period 1941–1951 in the form of a table with 11 rows and 2 columns (tutorial steps 13 and 14). This data can be entered as a *lookup table* named HTABLE by following these directions (which also appear in the tutorial):

- Start by clicking on the **Equations** entry on the menu bar and choosing **Lookup Tables** to display the lookup table manager window.
- Double click on <**Create New Table**> to display the new table window. Enter the name HTABLE, and specify 11 rows and 2 columns in the appropriate boxes. Then click the **OK** button. An array of empty cells will appear with 11 rows and 2 columns.
- Enter the data (provided in the tutorial window) in the array by clicking on cell [1, 1] to start. When all of the data is entered, click the **OK** button.
- Close the lookup table manager window.

Now you have a lookup table called HTABLE.

☞ Defining the harvest.

Go to the equation quadrant and, on a new line, add the following

$$harvest = \text{lookupval}(\text{HTABLE}, 1, t, 2)$$

Be sure to click the **Enter** box. The value returned by lookupval is the data in column 2 of HTABLE corresponding to the *t*-value of the data in column 1. (This value is computed by linear interpolation.)

☞ Using the **2D** custom tab.

Let's now look at this harvest data. Since *harvest* is not an ODE state variable, the Architect does not automatically generate a plot tab; we will have to make it by hand. Click on the **2D** tab at the lower right to select what we want to plot on each of the two axes. Place the cursor on the lower right plot quadrant, and after clicking the *right-most* mouse button, select **Edit** with the *left-most* mouse button. Leave the **X-Axis** variable set to **t**. For the **Y-Axis** click on the down arrow after **1.** **<None>** and select **harvest**. Now click on **Titles** at the top of the edit window and type in the **Graph Title** box: Harvest. In the **X-axis Title** box type: Year; and in the **Y-axis Title** box: Harvest. See Figure 1.10. Click **OK**. Using the *right-most* mouse button again on the lower graph, select **Scales**. Set the **X-Scale** as follows: deselect **Auto Scale**; set **Minimum** = 1940; **Maximum** = 1955; **Number of Ticks** = 3; **Label every** = 2. Select the **Y-Scale**, deselect **Auto Scale**, and set: **Minimum** = 0; **Maximum** = 1; **Number of Ticks** = 5; and **Label every** = 1. Click **OK**.

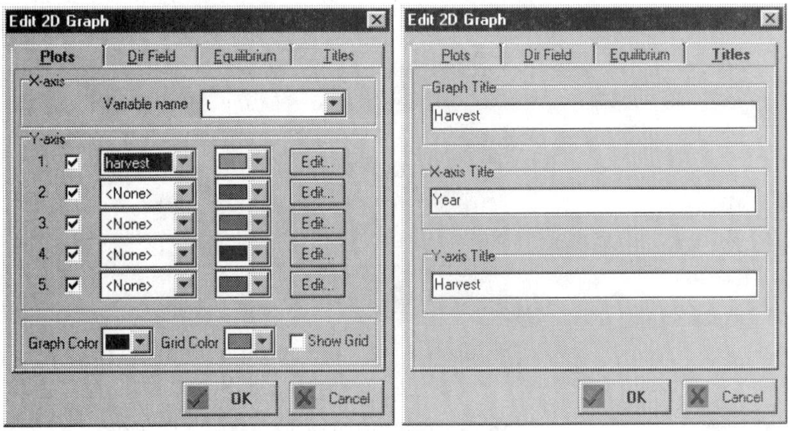

Figure 1.10: **Plots** and **Titles** panels for **2D** tab.

Click on the **Solve** icon. Notice that a graphical representation of the Pacific sardine landings appears in the lower graph but the upper graph has not been affected. That's because we have not included harvest (landings) in the sardine model yet. (Note: the two graphs have different vertical axis scales.)

☞ Including *harvest* in the sardine model.

After clicking on **Clear** and choosing **Clear All Runs**, go to the equation quadrant and modify the *sardine* ODE as follows:

$$sardine' = r * sardine * (6 - sardine)/6 - harvest \qquad (5)$$

You may have to scroll the equations quadrant (on some computers) in order to see the whole equation. (You can also move the dividing line between the right and left quadrants, at the slight expense of the graphing resolution.) Click the **Enter** box. Before we run the model, we must change the initial conditions to reflect the reality of the Pacific sardine population at that time. In the literature, the most reliable data for the Pacific sardine biomass starts in 1941. Thus set the **IC** for *t* to 1941 and the **IC** for *sardine* to 2.71. Reset **Interval** to 10. Now click the **Solve** icon and note the results. For best viewing of the top right graph window choose the **X-Scale**; deselect **Auto Scale**; set

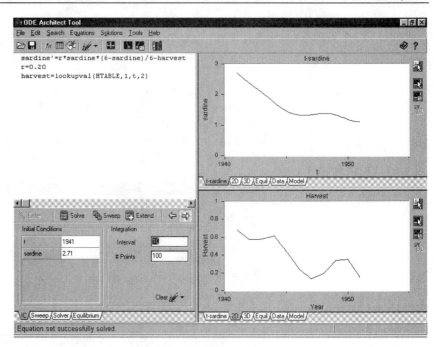

Figure 1.11: Model sardine biomass (upper), actual sardine biomass (lower).

Minimum = 1940; **Maximum** = 1955; **Number of Ticks** = 3; **Label every** = 2. Rescale the **Y-Scale** axis to: **Minimum** = 0; **Maximum** = 3; **Number of Ticks** = 3; **Label every** = 1. See Figure 1.11 for the graphs.

☞ Fishery collapse.

How do these model results compare with the expected behavior at the beginning of the chapter? While the overall behavior is captured in general terms by the model, it is unusual to have a model match the estimated data exactly.

✓ What are your thoughts about the model as it relates to historical behavior? Explain any discrepancies.

Step 8: Interpret the implications of the model

☞ Analysis.

It is now clear that while over-exploitation of the sardine landings was not the sole factor, it played a very large role in the collapse of the California fishery in the early 1950s. Since we now have a functioning model of that ten-year period in time, you have the amazing power to use your computer to revise history and attempt to save the fishing industry. What limit on the landings would have allowed a sizable sardine harvest[1] but not a collapse of the fishery?

---

[1]Historical note: A limit to the total catch of sardines at between 200,000 and 300,000 tons was recommended as early as 1929, and repeatedly over the next several years.

Now it is your turn to examine some options and try some alternate scenarios in the Explorations that follow.

◆ How to Model in Eight Steps

Modeling a situation mathematically involves many ideas and activities, but modeling is not always straightforward. There are many times when you may be puzzled, confused, and frustrated and you must retrace or rethink the steps involved. We summarize the steps in an order that allows for easy reference, but keep in mind the need to retreat, reassess, and redefine your thinking.

1. State the problem and its context.
2. Identify and assign variables.
3. State the laws that govern the relationships between the variables.
4. Translate the laws into equations.
5. Solve the resulting equations.
6. Interpret and test the solutions in the context of the natural environment.
7. Refine the model until it predicts the empirical data.
8. Interpret the implications of the model.

References   Barnes, J.T., MacCall, A., Jacobson, L.D., Wolf, P., "Recent population trends and abundance estimates for the Pacific sardine (Sardinops Sagax)" in *Calif. Coop. Oceanic Fish. Invest. Rep.*, Vol. 33 (1992), pp. 60–75

Cushing, D.H., "The problems of stock and recruitment" in *Fish Population Dynamics*, J.A. Gulland, Ed. (1977: J. Wiley & Sons, Inc.), pp. 116–133

Hocutt, C.H., and Stauffer, J.R., Eds., *Biological Monitoring of Fish* (1980: Lexington Books, D.C. Heath and Co.)

Lluch-Belda, D., Hernandez-Vazquez, S., Schwartzlose, R.A., "A hypothetical model for the fluctuation of the California sardine population" in *Long-term Variability of Pelagic Fish Populations and Environment*, Kawasaki et al., Eds. (1991: Pergamon Press), pp. 293–300

Murphy, G.I., "Population biology of the Pacific sardine (Sardinops Caerulea)" *Proc. Calif. Acad. Sci., Fourth Series* Vol. 34 (1966) No. 1, pp. 1–84

Radovich, J., "The collapse of the California sardine fishery. What have we learned?" in *Resource Management and Environmental Uncertainty*, M.H. Glantz, Ed. (1981: J. Wiley & Sons, Inc.), pp. 107–136

Ricker, W.E., "Computation and interpretation of biological statistics of fish populations" in *Bulletin 191* (1975: Department of the Environment Fisheries and Marine Service, Ottawa)

Wolf, P., "Recovery of the Pacific sardine and the California sardine fishery" in *Calif. Coop. Oceanic Fish. Invest. Rep.*, Vol. 33 (1992), pp. 76–86

Answer questions in the space provided, or on
attached sheets with carefully labeled graphs. A
notepad report using the Architect is OK, too.

Name/Date _____

Course/Section _____

Exploration 1.1. Constant Harvesting of a Biomass

1. *No harvesting.*
   Let's examine the rate of growth (the derivative) of the sardine biomass using the logistic model of ODE (3). To do this we'll look at the values of *sardine'* as a function of sardine biomass size. Go to the equations quadrant and type in the ODE

   $$sardine' = r * sardine * (6 - sardine)/6$$
   $$r = 0.20$$

   Click the **Enter** box.

   To create a plot of *sardine* vs. *sardine'*, select the **2D** tab (if necessary), place the cursor over the lower right graph, press the *right-most* mouse button, and select **Edit**. For the **X-Axis** use the down arrow to select *sardine*. For **Y-Axis 1**, select *sardine'*. Click now on the **Titles** tab at the top of the edit window. Type Rate of Growth vs. Biomass as the Graph Title, Sardine as the X-axis label and Sardine' as the Y-axis label. Click **OK**. Place the cursor over the lower right graph again, press the *right-most* mouse button and select **Auto Scales: Both** (if necessary). Next set the **IC** for *sardine* to 1. Click **Clear** and select **Clear All Runs** (if necessary), then click the **Solve** icon.

   The top graph shows (by default) *sardine* vs. time. Notice in the lower graph that the sardine growth rate, *sardine'*, is maximized somewhere near a midsized sardine population of about 3 million tons. Rescale the Y-axis of the top graph (if necessary) to **Minimum = 0; Maximum = 6; Number of Ticks = 6**. Verify that the sardine biomass grows at the rate of approximately 10% to 40% per year, depending upon the size of the biomass.

2. *Constant harvesting.*

Let's analyze the effect of constant harvesting on the logistic sardine popula-tion of Problem 1. Since the sardine biomass was 2.71 million tons in 1941, reset the *sardine* IC to 2.71 and keep the *sardine* vertical scale set on the range 0 to 6 and re-solve to observe the relative stabilization of the population.

Now insert a constant harvesting term in the model by modifying the ODE in the equation quadrant to read

$$sardine' = r * sardine * (6 - sardine)/6 - harvest$$

Try a harvest value that is slightly less than the biomass growth amount for 2.71 million tons by setting a constant harvest in the equation quadrant. For example you could try *harvest* = 0.28 (280,000 tons per year) and solve the model. (Be sure to click on the **Enter** box first.)

☞ Clearing a 2D custom graph. The same procedure clears a 3D custom graph.

Now click on the **2D** tab in the lower graph quadrant. Clear the graph in that quadrant by setting all axes to <**None**> in the **Plots** tab of the **Edit** box, then going to the **Titles** tab and deleting all titles. For the upper right graph you can set up and run a sweep of *harvest* over the values 0.1 (100,000 tons/year) to 0.7 (700,000 tons/year) using 7 points in the sweep. Describe the biomass behavior for harvest levels of 0.1; 0.3; 0.5; 0.7. From your explo-ration, determine what constant harvest amount provides a large harvest yet does not jeopardize the long-term viability of the Pacific sardine population. Did the harvest levels suggested by fishery researchers stand up?

3. *How does the IC affect the optimum harvest level?*

Is the optimum harvest level that you determined in Problem 2 affected by the initial biomass of the sardine in 1941? Try some different values for the IC and explain what you learn about the relationship between initial biomass and the optimum constant harvest amount.

Answer questions in the space provided, or on
attached sheets with carefully labeled graphs. A
notepad report using the Architect is OK, too.

Name/Date _____

Course/Section _____

Exploration 1.2. Constant Effort Harvesting

1. *Using a constant effort harvesting function.*
   Another model for harvesting is to land a certain percentage of the existing biomass each year. This is called *constant effort harvesting*. Introduce constant effort harvesting into ODE (5) by setting

   $$harvest = 0.25 * sardine$$

   to harvest 25% of the sardine population each year. Try a run. What happens? Go back and revise the harvest function to

   $$harvest = k * sardine$$

   and sweep through several values of your choosing for the harvest percentage k. Summarize your results. What is the optimum harvest percentage?

2. *How does the IC affect the optimum harvest percentage?*
Run some experiments to determine if the optimum harvest percentage you
select in Problem 1 is sensitive to the initial biomass of the sardine in 1941.
Explain your results. How do your results compare to the results of Problem 3
in Exploration 1.1?

Answer questions in the space provided, or on attached sheets with carefully labeled graphs. A notepad report using the Architect is OK, too.

Name/Date _____

Course/Section _____

Exploration 1.3. Investigating a Harvesting Function

1. *A unifying harvest strategy.*

    We can combine the strategies used in Explorations 1.1 (Problem 2) and 1.2 (Problem 1) by using a function that approximates each strategy at the appropriate time: proportional harvest for small sardine biomass and constant harvest for sufficiently large sardine biomass. A function suitable to this purpose is

    $$harvest = \frac{\alpha * sardine}{\beta + sardine}$$

    Use some algebra to demonstrate that the function does behave as claimed. Approximately what is the proportion? Approximately what is the constant harvest level?

2. *Testing the function.*

Determine values for α and β suitable for the Pacific sardine based on what you learned from Explorations 1.1 (Problem 2) and 1.2 (Problem 1). Is the optimal choice of α and β dependent on the initial biomass of the sardine?

Answer questions in the space provided, or on attached sheets with carefully labeled graphs. A notepad report using the Architect is OK, too.

Name/Date ⎯⎯⎯⎯⎯⎯⎯⎯⎯⎯⎯⎯

Course/Section ⎯⎯⎯⎯⎯⎯⎯⎯⎯⎯

Exploration 1.4. The Ricker Growth Rate Model

☞ Compare the Ricker with the logistic function: $R = rP(1 - P/K)$ for positive constants r and K.

Biologists commonly use the *Ricker function* to model fish population reproduction. The Ricker function is $R = \alpha P e^{(P_r - P)/P_m}$, where R is the reproduction rate, α is a constant, P is the parental or spawning stock population, P_r is the stock size at which $R = P$, and P_m is the stock size that yields maximum reproduction in the absolute sense. Calibrated for the Pacific sardine during the time period 1941 through 1951, this function is: $R = 0.15 P e^{(2.4 - P)/1.7}$.

1. *The Ricker population model.*
   Replace the logistic term in ODE (5) with the Ricker function to obtain

   $$sardine' = 0.15 * (sardine) * \exp((2.4 - sardine)/1.7)$$

   This function exhibits "compensatory behavior" that biologists know many fish populations exhibit. Plot two sardine populations vs. time on the same set of axes for comparison: *sardine*1′ as per the Ricker function above and *sardine*2′ as per the logistic growth model used earlier. You have to select the **2D** tab on the graphics window when defining the graph to get both populations on the same graph. To compare their respective growth patterns, plot the two sardine populations from 1920 to 1960 (**# Points** = 40) with **IC** set to 1 on both plots and with no harvesting. Based on this comparison, speculate what "compensatory" behavior is as envisioned by the biologists and reflected by the Ricker function.

2. *Repeat the harvest experiments.*

Repeat Exploration 1.1, Problem 2, using the Ricker function in the *sardine* ODE. What harvest level would provide a stable sustainable Pacific sardine population? Test whether the optimal harvest rate depends on the population IC. Are the results significantly different than when you used the logistic function?

2 Introduction to ODEs

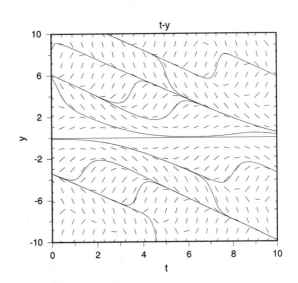

t-y

A slope field and some solution curves for $y' = y\sin(t + y)$.

Overview   Ordinary differential equations (ODEs) model many natural processes, so solutions of ODEs can be used to predict the behavior of those processes.

This chapter will investigate ODEs and initial value problems, their solutions, and their solution curves, along with some methods for finding solution formulas. Slope fields are introduced and used as guides to the behavior of solution curves. The path of a juggler's ball and the descent of a sky diver are modeled by ODEs.

Key words   Differential equation; solution; integration; separation of variables; initial values; modeling; slope field; direction field; juggling; sky diving; free fall; parachute; gravity; Newton's second law

See also   Chapter 1 for more on modeling, and Chapter 5 for more on models of motion.

◆ Differential Equations

Differential equations were first used in the seventeenth century to describe physical phenomena, such as the motion of orbiting planets or swinging pendulums. Since then they have been applied to processes, such as the growth of biological populations, the management of investment portfolios, and many other dynamical systems.

An *ordinary differential equation* is an equation involving an unknown function of one variable and one or more of its derivatives. For example, the ODE

$$\frac{dy}{dt} = y \cos t$$

is a statement about an unknown function y (the dependent variable) whose independent variable is t. To *solve* the ODE we need to find all the functions $y(t)$ that satisfy the ODE (we will discuss what we mean by a solution in the next section).

✓ "Check" your understanding by identifying the independent and dependent variables and the *order* of each ODE (i.e., the highest-order derivative that appears):

$$\frac{dy}{dx} = 2y + 2x$$

$$3\frac{d^2z}{dt^2} - 4\frac{dz}{dt} + 7z = 4\sin(2t)$$

◆ Solutions to Differential Equations

A function is a solution of an ODE if it yields a true statement when substituted into the equation. For example, $y = 2t^2$ is a solution of the equation

$$\frac{dy}{dt} = 4t \tag{1}$$

✓ Can you find another solution of ODE (1)?

☞ Most ODEs have infinitely many solutions.

Actually, ODE (1) has infinitely many solutions. A single solution is called a *particular solution*. The set of all solutions is called the *general solution*. For example, the general solution of ODE (1) is $y = 2t^2 + C$, where C is any constant, while $y = 2t^2 + 3$ is a particular solution.

◆ Solving a Differential Equation

Solving a differential equation involves finding a function, just as solving an algebraic equation involves finding a number.

An ODE such as $dy/dt = 2ty$ gives us information about an unknown function y in terms of its derivative(s). In your differential equations class, you'll learn some methods for finding solutions of ODEs. The section "Finding a Solution Formula" later in this chapter also describes two techniques.

◆ Slope Fields

☞ Slopes for $y' = y \cos t$:

Point	Slope
(0,0)	0
(0,1)	1
(0,2)	2
(0,−1)	−1
(0,−2)	−2
$(\frac{\pi}{2}, y)$	0

☞ Each segment of a slope field is tangent at its midpoint to the solution curve through that midpoint.

One useful way to get information about solutions of an ODE is to graph them; graphs of solutions are called *solution curves*. For first-order ODEs, you can actually get a good idea of what solution curves look like without solving the equation! Notice that for the ODE $y' = y \cos t$ the slope of the solution curve passing through the point (t, y) is given by $y \cos t$. Every first-order ODE gives you direct information about the slope of the solution curve through a point, so you can visualize solution curves by drawing small line segments with the correct slopes on a grid of fixed points. With patience (or a computer), you can draw many such line segments (as in the chapter cover figure). This is called a *slope field*. (Some books call it a *direction field*.) With practice you'll be able to imagine some of the line segments running together to make a graph. This approximates the graph of a solution to the ODE, that is, a solution curve. Figure 2.1 shows a slope field with several solution curves.

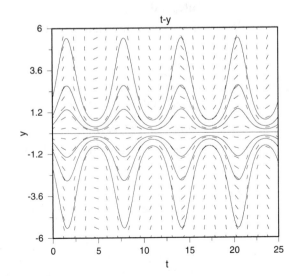

Figure 2.1: Slope field and seven solution curves for $y' = y \cos t$.

◆ Initial Values

We have seen that an ODE can have many solutions. In fact, the general solution formula involves an arbitrary constant. What happens if we specify that the solution must satisfy another property, such as passing through a given point? For example, all functions $y = 2t^2 + C$ are solutions of the ODE $dy/dt = 4t$, but only the *specific* solution $y = 2t^2 + 3$ satisfies the condition that $y = 5$ when $t = 1$. So, if we graph solution curves in the ty-plane, only the graph of the solution $y = 2t^2 + 3$ goes through the point $(1, 5)$.

The requirement that $y(1) = 5$ is an example of an *initial condition*, and the combination of the ODE and an initial condition

$$\frac{dy}{dt} = 4t, \quad y(1) = 5 \tag{2}$$

is called an *initial value problem* (IVP). Its solution is $y = 2t^2 + 3$.

✓ Replace the condition $y(1) = 5$ in IVP (2) by $y(2) = 3$ and find the solution of this new initial value problem. How many solutions are there?

◆ Finding a Solution Formula

An ODE usually has many solutions. How can you find a solution, and how can you describe it? A solution formula provides a useful description, but graphs and tables generated by ODE Architect are also useful, especially in the all-too-frequent case where no formula can be found. Two techniques to find solution formulas are summarized here, and others are in your textbook.

Integration

If $f(t)$ is a continuous function, then the general solution of the ODE

$$\frac{dy}{dt} = f(t)$$

☞ A table of integrals comes in handy here.

is $y(t) = F(t) + C$, where $F(t)$ is an antiderivative of f. For example, the general solution of $dy/dt = \sin t$ is $y = -\cos t + C$.

Separation of Variables

If you can write a differential equation in the form

$$\frac{dy}{dt} = f(t)g(y)$$

then wherever $g(y) \neq 0$ you can rewrite it as

$$\frac{1}{g(y)} \frac{dy}{dt} = f(t)$$

so that

$$\int \frac{1}{g(y)} \, dy = \int f(t) \, dt$$

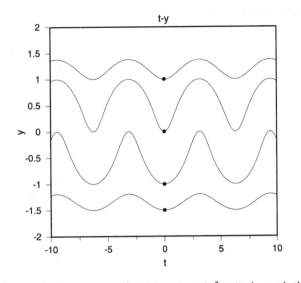

Figure 2.2: Four solution curves of $dy/dt = \sin t/(3y^2 + 1)$ through the marked initial points.

☞ Keep that table of integrals handy!

If $H(y)$ is an antiderivative of $1/g(y)$ and $F(t)$ is an antiderivative of $f(t)$, then a solution $y(t)$ of the ODE solves the equation

$$H(y(t)) = F(t) + C$$

for some constant C.

Here's an example of a separable ODE:

$$\frac{dy}{dt} = \frac{\sin t}{3y^2 + 1} \tag{3}$$

Separating the variables and finding the antiderivatives, we see that

$$(3y^2 + 1)\frac{dy}{dt} = \sin t$$

$$y^3 + y = -\cos t + C \tag{4}$$

We won't attempt to express a solution $y(t)$ directly in terms of t (and C), but we can check that formula (4) is correct by differentiating each side with respect to t. This gives

$$3y^2\frac{dy}{dt} + \frac{dy}{dt} = \sin t$$

which has the form of ODE (3) if we divide each side by $3y^2 + 1$. Figure 2.2 shows solution curves of ODE (3) through the initial points $(0, -1.5)$, $(0, -1)$, $(0, 0)$, $(0, 1)$. The curves were plotted by using ODE Architect to solve ODE (3) with the given initial data.

Solution formulas are useful, but they exist only for a small number of ODEs of special forms. That's where numerical solvers like ODE Architect come in—they don't need solution formulas.

◆ Modeling

☞ The eight steps are described in Chapter 1.

A *mathematical model* is a system of mathematical equations relating specific variables that represent some aspect of a natural process. Modeling involves several steps:

1. State the problem and its context.
2. Identify and assign variables.
3. State the laws that govern the relationships between the variables.
4. Translate the laws into equations.
5. Solve the resulting equations.
6. Interpret and test the solutions in the context of the natural environment.
7. Refine the model until it predicts the empirical data.
8. Interpret the implications of the model.

The models we consider all involve ODEs.

◆ The Juggler

You can observe the modeling process in the following juggler problem.

1. Find an ODE that describes the height of a ball between the time it leaves the juggler's hand, moving vertically upward, and the time it falls back into the hand.

2. Let t = time (in seconds), h = height of the ball above the floor (in feet), v = velocity (in ft/sec), and a = acceleration (in ft/sec^2).

3. Apply Newton's second law of motion to the ball: the mass m of a body times its acceleration is equal to the sum of all of the forces acting on the body. We treat the ball as a point mass encountering negligible air resistance (drag) so the only force acting on the ball is that due to gravity, which pulls the ball downward.

4. By Newton's second law, we have that $ma = -mg$, where $g = 32$ ft/sec^2 is the acceleration due to gravity near the surface of the earth, and the minus sign indicates the downward direction of the gravitational force. Since the ball's acceleration is $a = v'$ where v is its velocity, and $v = h'$, we can model the ball's motion by $h'' = -32$. The initial height h_0 of the ball is that of the juggler's hand above the floor when the ball is launched upward, and that is easy to measure. The initial velocity v_0 is harder to measure directly; it is simpler to solve the model first and then experiment to deduce a reasonable value for v_0.

☞ So the juggler's ODE for vertical motion is $h'' = -32$.

5–8. Solving and testing are up to you. See Figure 2.3 for graphs of $h(t)$ corresponding to $h_0 = 4$ ft and five values of v_0.

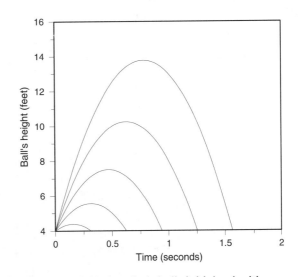

Figure 2.3: Five tosses of the juggler's ball: initial velocities v_0 range from 5 to 25 ft/sec. Which time-height curve corresponds to $v_0 = 25$?

✓   How must you revise the process when the ball is thrown to the juggler's *other* hand? (The result appears on Screen 3.4 of Module 2.)

◆ The Sky Diver

You might think that the path of a sky diver in free fall looks like the downward path of the ball in the simplest juggler problem of vertical motion. However, as the sky diver's velocity becomes large the effects of air resistance (or *drag*) become noticeable and must be included in the model. A revised model (starting with Step 3) follows:

☞ This kind of air resistance is called *viscous* damping.

3. In this case, Newton's second law says that mass times acceleration is equal to the force due to gravity plus that due to air resistance. Experience has shown that the force of air resistance can be modeled fairly well by a term that is proportional to velocity and opposite in direction.

4. We have $mh'' = mv' = -mg - kv$, where k is a constant coefficient of air resistance. The initial velocity of the sky diver is $v_0 = 0$ ft/sec; the initial height when the sky diver jumps from the plane is h_0 ft.

5. We solve the second-order ODE for h in two steps, first for v (by separating the variables) and then for h (by integrating the expression we

find for v, since $v = h'$). Here are the steps:

$$v' = -g - \frac{k}{m}v, \quad v(0) = 0$$

$$\frac{1}{g + kv/m}\frac{dv}{dt} = -1$$

$$\int \frac{1}{g + kv/m}\, dv = -\int dt$$

☞ *C* is an arbitrary constant.

$$(m/k)\ln(g + kv/m) = -t + C$$

$$\ln(g + kv/m) = (k/m)(-t + C)$$

Exponentiating and setting $K = \exp(kC/m)$ we obtain

$$g + kv/m = Ke^{-(k/m)t}$$

Since $v = 0$ when $t = 0$, we find that $K = g$. Solving for v we obtain

$$v = \frac{-mg}{k} + \frac{mg}{k}e^{-(k/m)t}$$

That means that $h(t)$ solves the IVP

$$h' = v = \frac{-mg}{k} + \frac{mg}{k}e^{-(k/m)t}, \quad h(0) = h_0$$

We find the formula for $h(t)$ by integration and the fact that $h = h_0$ at $t = 0$:

$$h = \frac{-mg}{k}t - \frac{m^2 g}{k^2}e^{-(k/m)t} + \frac{m^2 g}{k^2} + h_0$$

In our example of free fall (Screen 4.3), these equations become

☞ So the sky diver's free fall ODE is $h'' = -32 - (k/5)v$.

$$h'' = v' = -32 - \frac{k}{5}v \quad \text{if } m = 5 \text{ slugs}$$

$$h' = v = \frac{-160}{k} + \frac{160}{k}e^{-(k/5)t}$$

$$h = \frac{-160}{k}t - \frac{800}{k^2}e^{-(k/5)t} + \frac{800}{k^2} + 13500 \qquad (5)$$

See Figure 2.4 for some time-height curves.

Since the mass m of the sky diver doesn't drop out of the ODE when damping is added, we have to use appropriate units for the mass. In English units (which the English have been wise enough to discard) we have

$$\text{mass} = \frac{\text{force}}{\text{acceleration}} = \frac{\text{weight}}{\text{gravity}} = \frac{\text{lbs}}{\text{ft/sec}^2} = \text{slugs}$$

Opening the Parachute

If we wish to model what happens when the parachute opens, we'll need to alter the model slightly to account for the sudden change in drag—that is, for how the value of k suddenly changes.

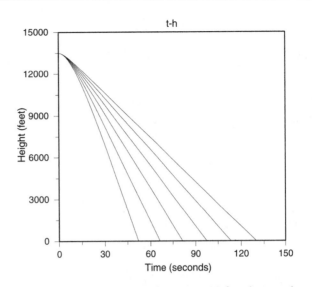

Figure 2.4: Six sky divers in free fall from 13,500 ft: viscous damping constants range from 0.5 to 1.5 slug/sec. Which sky diver has the smallest damping constant?

4. We can use experimental values for the drag coefficients: in free fall, $k_{ff} = 0.86$ and, after the parachute opens, $k_p = 6.71$, both in slugs/sec. The parachute opens at time t_p, when h is 2500 feet. It's hard to calculate t_p from formula (5) for h, so we can approximate it by reading the graph of h vs. t.

☞ Opening the chute changes k from k_{ff} to k_p.

We noticed on Screen 4.5 of Module 2 that an instantaneous opening of the parachute would exert an enormous force on the sky diver, so the model was further revised to allow the chute to open over a few seconds (a more realistic model), and we let k grow gradually, in a linear way, as it goes from k_{ff} to k_p. Take a look at Exploration 2.4, Problem 3.

References   Borrelli, R. L., and Coleman, C. S., *Differential Equations: A Modeling Perspective*, (1998: John Wiley & Sons, Inc.)

Boyce, W. E., and DiPrima, R. C., *Elementary Differential Equations and Boundary Value Problems*, 6th ed. (1997: John Wiley & Sons, Inc.)

Hale, M., and Skidmore, A., *A Guided Tour of Differential Equations*, (1997: Prentice-Hall)

C·ODE·E Newsletter, http://www.math.hmc.edu/codee, for articles on modeling with ODEs

IDEA (Internet Differential Equations Activities), created by Thomas LoFaro and Kevin Cooper, offers an interactive virtual lab book with models. http://www.sci.wsu.edu/idea

Answer questions in the space provided, or on
attached sheets with carefully labeled graphs. A
notepad report using the Architect is OK, too.

Name/Date _____

Course/Section _____

Exploration 2.1. ODEs and Their Solutions

1. *Where is that constant?*
   Solution formulas for first-order ODEs often involve an arbitrary constant C, and it can show up in all sorts of strange places in the formulas. Solve each of the following ODEs for y in terms of t and C.

   (a) $y' = 1 + \sin t$ **(b)** $y' = -y/3$ **(c)** $y' = t/y$ **(d)** $y' = 2t^2 y/\ln y$

2. *Let's check out the ODE Architect.*
   You can see how good the ODE Architect solver is by creating initial value problems for the ODEs of Problem 1 and using the Architect to solve them and graph the solutions. Then compare the solver graphs with those obtained using the solution formula. For example, use ODE Architect to solve and plot the solution of the IVP $y' = -y/3$, $y(0) = 1$. Then graph the solution $y = e^{-t/3}$ and compare. To do this, enter the following two equations on the editor screen:

   $$y' = -y/3$$
   $$u = e^{-t/3}$$

   Next enter the initial condition for the ODE, then solve and plot the solution on one of the graphics screens. Use the custom 2D plot tab to overlay the graph of u. Do the graphs match? Repeat with your own initial data for each of the other three ODEs in Problem 1.

3. *How many[1] solutions does this IVP have?*
Find formulas for two different solutions for the IVP $y' = y^{1/3}$, $y(0) = 0$.
Which solution does ODE Architect give? Repeat with $y' = y^{2/3}$, $y(0) = 0$.
[*Hint:* Is $y(t) = 0$ for all t a solution?]

4. *The effect of a singularity in the differential equation.*
The ODE $y' = y/t$ has a singularity at the point $(0, 0)$ because at that point,
$y/t = 0/0$, which is undefined. Find a formula for all solutions of the ODE.
Does the IVP $y' = y/t$, $y(0) = 0$, have any solutions? Use ODE Architect for
$y' = y/t$, $y(1) = a$, for various positive values of a and then solve backward
in time to see what happens as t gets near zero. Explain.

---

[1] Usually an IVP has a single solution, but in this Exploration you will see some exceptions. You
can find out why by reading about "existence" and "uniqueness" in your text.

Answer questions in the space provided, or on
attached sheets with carefully labeled graphs. A
notepad report using the Architect is OK, too.

Name/Date _____

Course/Section _____

Exploration 2.2. Slope Fields

1. *What happens in the long term?*
   The following ODEs are given in Screen 2.2 (Experiment 1). Using ODE Architect, describe what the solutions do as t gets very large. Include sketches or printouts of your solution curves and their slope fields.

   (a) $y' = y - 1$ **(b)** $y' = t/4$ **(c)** $y' = (y - t)/10$

2. *More long-term behavior.*
   Repeat Problem 1 with the following ODEs.

   (a) $y' = ty$ **(b)** $y' = (y^2 - 4)/10$ **(c)** $y' = (y - 3)/5$

3. *Still more long-term behavior.*
    Using ODE Architect, describe the long-term behavior of the solutions of
    $y' = y \sin(t + y)$.

4. *Strange solutions.*
    Make up your own ODEs, especially ones whose solution curves or slope
    fields form strange patterns. Use ODE Architect to display your results. De-
    scribe the long-term behavior of solution curves. Attach printouts of your
    graphs.

Answer questions in the space provided, or on
attached sheets with carefully labeled graphs. A
notepad report using the Architect is OK, too.

Name/Date _____

Course/Section _____

Exploration 2.3. The Juggler

Second-order ODEs of the form $y'' = f(t, y, y')$ are to be solved in Explorations 2.3 and 2.4. Since ODE Architect only accepts first-order ODEs, we will replace $y'' = f$ by an equivalent pair of first-order ODEs. We do this by introducing $v = y'$ as another dependent variable:

$$y' = v$$
$$v' = f(t, y, v)$$

1. *What goes up must come down.*
   Use ODE Architect to find the position of the ball at several different times t for several different initial velocities. Assume no air resistance and that the ball moves in a vertical line. What is the name for the shapes of the solution curves in the ty-plane? Does it take longer for the ball to rise or to fall? Show and explain the difference (if there is one!).

2. *Hand-to-hand motion of the ball.*
   For a given initial speed v_0, find the range of values of the angle θ_0 so that the ball goes from one hand to the other. Now increase the initial speed. What happens to the range of successful values of θ_0? Explain. [Suggestion: First take a look at Screen 3.5 (Experiment 2 in Module 2); then enter the equations in ODE Architect and vary θ_0 with fixed v_0 to find the ranges. You may also want to take a look at Screens 1.2 and 1.3 in Module 5.]

3. *Raise your hand!*

   Suppose the juggler raises his catching hand one foot higher. Repeat Problem 2 in this setting.

4. *Juggling two balls.*

   Construct model ODEs for tossing two balls, one after the other, from one hand to the other. Use ODE Architect to find the positions of both balls at time t.

Answer questions in the space provided, or on
attached sheets with carefully labeled graphs. A
notepad report using the Architect is OK, too.

Name/Date _____

Course/Section _____

Exploration 2.4. The Sky Diver

Second-order ODEs of the form $y'' = f(t, y, y')$ are to be solved in Explorations 2.3 and 2.4. Since ODE Architect only accepts first-order ODEs, we will replace $y'' = f$ by an equivalent pair of first-order ODEs. We do this by introducing $v = y'$ as another dependent variable:

$$y' = v$$
$$v' = f(t, y, v)$$

1. *Terminal speed of a falling body.*
   Use ODE Architect and determine the sky diver's terminal speeds for several different values of the viscous damping coefficient (use $m = 5$ slugs and $g = 32$ ft/sec^2). Is there any difference if the sky diver jumps at 25,000 feet instead of 13,500 feet? [*Suggestion:* After entering the ODE and solving, click on a Data tab in either of the two graphics windows and use approximate data you find there.]

2. *Slow down!*
   If a sky diver can survive a free-fall jump only if she hits the ground at no more than 30 ft/sec, what values of the viscous drag coefficient k make this possible? Are these k-values realistic? (Use $m = 5$ slugs and $g = 32$ ft/sec^2.)

3. *A Modeling Challenge!*

   Let's construct a model for a parachute that opens over a 3 second time span. The ODEs for this model are given on Screen 4.5 (Experiment 2), but we have to define $k(t)$. Assume that the sky diver has a mass of 5 slugs and that she jumps from 13,500 ft. The parachute starts to open after 65 seconds of free fall and the damping coefficient changes linearly from $k_{ff} = 0.86$ slugs/ft to $k_p = 6.71$ slugs/ft as the chute opens. In other words,

   $$k(t) = \begin{cases} k_{ff}, & t < 65 \\ k_{ff} + \frac{k_p - k_{ff}}{3}(t - 65), & 65 \leq t \leq 68 \\ k_p, & t > 68 \end{cases}$$

☞ A step function is one of the engineering functions. You can find them by going to ODE Architect and clicking on Help, Topic Search, and Engineering Functions.

   (a) Write an expression for $k(t)$ using the properties of step functions. *Hint*:

   $$\text{Step}(t, 65) - \text{Step}(t, 68) = \begin{cases} 1, & 65 \leq t \leq 68 \\ 0, & \text{otherwise} \end{cases}$$

   (b) Use ODE Architect to plot the sky diver's acceleration, velocity, and height vs. time, using your expression for $k(t)$.

3 Some Cool ODEs

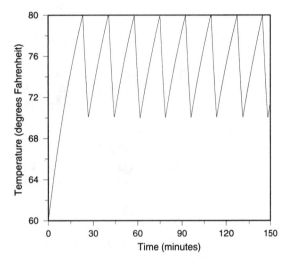

A room heats up in the morning, and the air conditioner in the room starts its on-off cycles.

Overview   In this chapter, we'll use Newton's law of cooling to build mathematical models of a number of situations that involve the variation of temperature in a body with time. Some of our models involve ODEs that can be solved analytically; others will be solved numerically by ODE Architect. We'll compare the analytical solutions and the numerical results and see how both can be used to verify predictions made by the models.

Key words   Modeling; Newton's law of cooling (and warming); initial conditions; general solution; separation of variables; integrating factor; heat energy; melting; air conditioning

See also   Chapter 1 for more on modeling and Chapter 2 for the technique of separation of variables.

◆ Newton's Law of Cooling

Have you ever gotten an order of piping hot French fries, only to find them ice cold in what seems like a matter of moments? Whenever an object (or substance) is warmer than its surroundings, it cools because it loses heat energy. The greater the temperature difference between the object and its surroundings, the faster the object cools. The temperature of a body rises if its surroundings are at a higher temperature than it is. What happens to the ice cream in a cone on a hot day?

Although it is an oversimplification, we will assume that the temperature is uniform at all points in the objects we wish to model, but the temperature may change with time. Let's assume that the rate of change of the object's temperature is proportional to the difference between its temperature and that of its surroundings. Stated mathematically, we have:

☞ This becomes a "law of warming" if the surroundings are hotter than the object.

> **Newton's law of cooling.** If $T(t)$ is the temperature of an object at time t and $T_{out}(t)$ is the temperature of its surroundings, then
>
> $$\frac{dT}{dt} = k(T_{out} - T) \qquad (1)$$
>
> where k is a positive constant called the *cooling coefficient*.

◆ Cooling an Egg

What happens to the temperature of a hard-boiled egg when you take it out of a pot of boiling water? At first, the egg is the same temperature as the boiling water. Once you take it out of the water the egg begins to cool, rapidly at first and then more slowly. The temperature of the egg, $T(t)$, drops at a rate proportional to the difference between the temperature of the air, T_{out}, and $T(t)$. Notice from ODE (1) that if $T_{out} < T(t)$, the rate of change of temperature, dT/dt, is negative, so $T(t)$ decreases and your egg cools.

✓ "Check" your understanding by answering this question: What happens to the temperature of an egg if it is boiled at 212°F and then transferred to an oven at 400°F?

◆ Finding a General Solution

Equation (1) is a first-order ODE and its general solution contains one arbitrary constant. We can see this as follows: If T_{out} is a constant, then ODE (1) is separable, and separating the variables we have

☞ See Chapter 2 for how to solve a separable ODE.

$$\int \frac{dT}{T_{out} - T} = \int k\, dt$$

Finding an antiderivative for each side we obtain

☞ Why are the absolute value signs needed?

$$-\ln|T_{out} - T(t)| = kt + K$$

where K is an arbitrary constant. Multiplying through by -1 and exponentiating gives us

$$|T_{out} - T(t)| = e^{-K}e^{-kt}$$

or, after dropping the absolute value signs, we have that

$$T(t) = T_{out} + Ce^{-kt} \qquad (2)$$

where $C = \pm e^{-K}$ is now the arbitrary constant. The solution formula (2) is called the *general solution* of ODE (1).

✓ How does the temperature $T(t)$ in (2) behave as $t \to +\infty$? Why can the constant C be positive or negative?

Given an initial condition, we can determine C uniquely and identify a single solution from the general solution (2). If we take $T(0) = T_0$, then since $T(0) = T_{out} + C$ we see that $C = T_0 - T_{out}$ and we get the unique solution

$$T(t) = T_{out} + (T_0 - T_{out})e^{-kt} \qquad (3)$$

The constant of proportionality, k, in ODE (1) determines the rate at which the body cools. It can be evaluated in a number of ways, for example, by measuring the body's temperature at two different times and using formula (3) to solve for T_0 and k. Figure 3.1 shows temperature curves corresponding to $T_{out} = 70°F$, $T_0 = 212°F$, and five values of k.

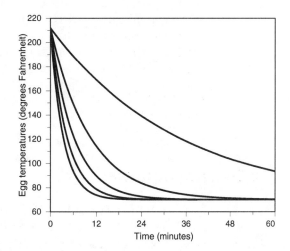

Figure 3.1: The cooling coefficient k ranges from 0.03 to 0.3 min^{-1} for eggs of different sizes. Which is the $k = 0.03$ egg?

✓   An object is initially at 212°F and cools to 190°F after 5 minutes in a room that is at 72°F. Find the coefficient of cooling, k, and determine how long it will take to cool to 100°F.

Finding the general solution formula (2) for ODE (1) was straightforward. However, the vast majority of ODEs are not so simple to solve and we have to use numerical methods. To demonstrate the accuracy of such methods, you can compare the numerical solutions from ODE Architect with a known solution formula.

☞ ODE Architect helps out again.

✓   How long will it take for a 212°F egg to cool to 190°F in a 72°F room if $k = 0.03419$ min^{-1}? Use ODE Architect and formula (3) and compare the results.

◆ Time-Dependent Outside Temperature

When considering the cooling of an egg, ODE (1) is separable because T_{out} is constant in this instance. Let's consider what happens when the outside temperature changes with time.

We can still use Newton's law of cooling, so that if $T(t)$ is the egg's temperature and $T_{out}(t)$ is the room's temperature, then

$$\frac{dT}{dt} = k(T_{out}(t) - T) \qquad (4)$$

Note that ODE (4) is not separable (because T_{out} varies with time) but it is linear, so we can find its general solution as follows. Rearrange the terms to give the linear ODE in standard form:

$$\frac{dT}{dt} + kT = kT_{out}(t)$$

Multiply both sides by e^{kt}, so that

$$e^{kt}\left(\frac{dT}{dt} + kT\right) = kT_{out}(t)e^{kt} \qquad (5)$$

Since the left-hand side of ODE (5) is $(d/dt)(e^{kt}T(t))$, it can be rewritten:

$$\frac{d}{dt}\left(e^{kt}T\right) = kT_{out}(t)e^{kt} \qquad (6)$$

Integrating both sides of ODE (6) we have that

$$e^{kt}T = \int kT_{out}(t)e^{kt}dt + C$$

where C is an arbitrary constant. The magic factor $\mu(t) = e^{kt}$ that enabled us to do this is called an *integrating factor*. So ODE (4) has the general solution

☞ Every ODE text discusses integrating factors and first-order linear ODEs.

$$T(t) = e^{-kt}\left(\int kT_{out}(t)e^{kt}dt + C\right)$$

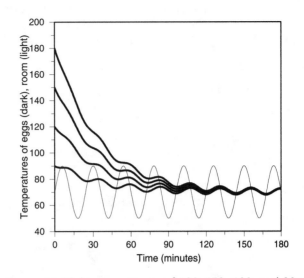

Figure 3.2: Eggs at initial temperatures of 180, 150, 120, and 90°F cool in a room whose temperature oscillates sinusoidally about 70°F for $k = 0.03$ min^{-1}. Do the initial temperatures matter in the long term?

Finally, letting $T(0) = T_0$ and integrating from 0 to t, we get the solution

$$T(t) = e^{-kt}\left(\int_0^t kT_{out}(s)e^{ks}\,ds + T_0\right) \tag{7}$$

It may be possible to evaluate the integral (7) analytically, but it is easier to use ODE Architect right from the start. See Figure 3.2 for egg temperatures in a room whose temperature oscillates between hot and cold.

✓   Show that if T_{out} is a constant, then formula (7) reduces to formula (3).

☞ You may find a computer algebra system or a table of integrals helpful!

✓   Use equation (7) to find a formula for $T(t)$ if

$$T_{out}(t) = 82 - 10\sin\left(\frac{2\pi(t+3)}{24}\right)$$

Use a table of integrals to carry out the integration.

◆ Air Conditioning a Room

Now let's build a model that describes a room cooled by an air conditioner. Without air conditioning, we can model the change in temperature using ODE (1). When the air conditioner is running, its coils remove heat energy at a rate proportional to the difference between T_r, the room temperature, and the temperature T_{ac} of the coils. So, using Newton's law of cooling for the temperature change due to both the air outside the room and the air conditioner coils, our model ODE is

☞ Newton's law of cooling (twice)!

$$\frac{dT_r}{dt} = k(T_{out} - T_r) + k_{ac}(T_{ac} - T_r)$$

where T_{out} is the temperature of the outside air and k and k_{ac} are the appropriate cooling coefficients. If the unit is turned off, then $k_{ac} = 0$ and this equation reduces to ODE (1).

Let's assume that the initial temperature of the room is 60°F and the outside temperature is a constant 100°F. The air conditioner operates with a coil temperature of 40°F, switches on when the room reaches 80°F, and switches off at 70°F. Initially, the unit is off and the change in the room temperature is modeled by

☞ Time *t* is measured in minutes.

$$\frac{dT_r}{dt} = 0.03(100 - T_r), \quad T_r(0) = 60 \tag{8}$$

where we have taken the cooling coefficient $k = 0.03 \text{ min}^{-1}$. As we expect, the temperature in the room will rise as time passes. At some time t_{on} the room's temperature will reach 80°F and the air conditioner will switch on. If $k_{ac} = 0.1 \text{ min}^{-1}$, then for $t > t_{on}$ the temperature is modeled by the IVP

$$\frac{dT_r}{dt} = 0.03(100 - T_r) + 0.1(40 - T_r), \quad T_r(t_{on}) = 80 \tag{9}$$

which is valid until the room cools to 70°F at some time t_{off}. Then for $t > t_{off}$ the room temperature satisfies the IVP (8) but with the new initial condition $T_r(t_{off}) = 70$. Each time the unit turns on or off the ODE alternates between the two forms given in (8) and (9).

☞ The modeling here is more advanced than you have seen up to this point. You may want to just use the equations and skim the modeling.

Solving the problem by hand in the manner just described is very tedious. However, we can use ODE Architect to change the ODE automatically and without having to find t_{on} and t_{off}. The key is to define k_{ac} to be a function of temperature by using a step function; here's how we do it. In the equation quadrant of ODE Architect write the ODE as

☞ The step function is one of the engineering functions. You can find them by going to ODE Architect and clicking on Help, Topic Search, and Engineering Functions.

$$Tr' = 0.03 * (100 - Tr) + kac * (40 - Tr)$$

Now define k_{ac} as follows:

$$kac = 0.1 * \text{Step}(Tr, Tc)$$

where

$$Tc = 75 + 5 * B$$

Here T_c is the control temperature and

$$B = 2 * \text{Step}(Tr', 0) - 1$$

Note that $B = +1$ when $T_r' > 0$ (the room is warming) and $B = -1$ when $T_r' < 0$ (the room is cooling). This causes T_c to change from 80°F to 70°F (or

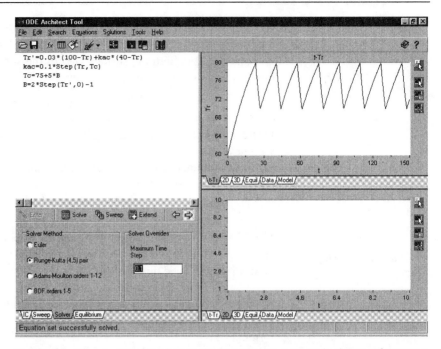

Figure 3.3: Air conditioning keeps the room temperature in the comfort zone, $70°F \le T_r \le 80°F$.

the reverse) depending on whether the room is warming or cooling. Finally, k_{ac} is zero (the air conditioner is off) when $T_r < T_c$, and $k_{ac} = 0.1$ (the air conditioner is on) when $T_r > T_c$.

The overall effect is that the air conditioner switches on only if the room temperature is above 80°F, then it runs until the room is cooled to 70°F, and then it switches off. The room temperature rises again to 80°F, and the process repeats. The temperature-vs.-time plot is shown in Figure 3.3. The accompanying screen image shows that we have set the maximum time step to 0.1 (under the Solver tab). If the internal time steps are not kept small, the Architect will not correctly notice when the step functions turn on and off.

◆ The Case of the Melting Snowman

It is difficult to model the melting of a snowman because of its complicated geometry: a large roundish ball of snow with another smaller mound on top. So let's simplify the model by treating the snowman as a single spherical ball of snow. The rate at which the snowman melts is proportional to the rate at which it gains thermal energy from the surrounding air, and it is given by

$$\frac{dV}{dt} = -h\frac{dE}{dt} \tag{10}$$

where V is the ball's volume, E is thermal energy, and h is a positive constant.

Our snowman will gain thermal energy only at its surface, where it is exposed to the warm air. So, it is reasonable to assume that the energy gain is proportional to both the surface area of the snowman and the temperature difference between the air and the snow:

☞ Remember that the snowman's temperature is always 32°F.

$$\frac{dE}{dt} = \kappa A(V)(T_{out} - 32) \tag{11}$$

where κ is a positive constant, and $A(V)$ is the surface area of a sphere of volume V.

☞ This is the snowman's law of melting.

If we combine equations (10) and (11) and take $k = \kappa h$, we obtain

$$\frac{dV}{dt} = -kA(V)(T_{out} - 32) \tag{12}$$

✓ The volume of a sphere of radius r is $V = \frac{4}{3}\pi r^3$ and its surface area is $A = 4\pi r^2$. Eliminate r between these two formulas to express A as a function of V. (You will need this soon.)

Note that ODE (12) is separable even when the outside temperature T_{out} is a function of time. Separating the variables and integrating we find the formula

$$\int \frac{1}{A(V)} dV = -\int k(T_{out}(t) - 32)\, dt + C \tag{13}$$

which defines V implicitly as a function of t. We can find the constant of integration C from the volume of the snowman at a specific time. However, expressions for the integrals in formula (13) may be hard to find. Once again ODE Architect comes to the rescue and solves ODE (12) numerically, given formulas for $A(V)$, $T_{out}(t)$, and the initial volume.

✓ If $k = 0.1451$ ft/(hr °F), the original volume of the snowman is 10 ft³, and the outside temperature is 40°F, how many hours does it take the snowman's volume to shrink to 5 ft³?

References   Nagle, R.K., and Saff, E.B., *Fundamentals of Differential Equations*, 3rd ed. (1993: Addison-Wesley)

Farlow, S.J., *An Introduction to Differential Equations and their Applications*, (1994: McGraw-Hill)

Answer questions in the space provided, or on
attached sheets with carefully labeled graphs. A
notepad report using the Architect is OK, too.

Name/Date _____

Course/Section _____

Exploration 3.1. Cooling Bodies

1. *Too hot to handle.*
   When eating an egg, you don't want it to be too hot! If an egg with an initial temperature of 15°C is boiled and reaches 95°C after 5 minutes, how long will you have to wait until it cools to 70°C?

2. *A dead body, methinks.*
   In forensic science, it is important to be able to estimate the time of death if the circumstances are suspicious. Assume that a corpse cools according to Newton's law of cooling. Suppose the victim has a temperature of 72°F when it is found in a 40°F walk-in refrigerator. However, it has cooled to 66.8°F two hours later when the forensic pathologist arrives. Estimate the time of death.[1]

[1] From "Estimating the Time of Death" by T.K. Marshall and F.E. Hoare, *Journal of Forensic Sciences*, Jan. 1962.

3. *In hot water.*

    Heat a pan of water to 120°F and measure its temperature at five-minute in-
    tervals as it cools. Plot a graph of temperature vs. time. For various values
    of the constant k in Newton's law of cooling, use ODE Architect to solve the
    rate equation for the water temperature. What value of k gives you a graph
    that most closely fits your experimental data?

4. *More hot water.*

    In Problem 3 you may have found it difficult to find a suitable value of k. Here
    is the preferred way to determine k. The solution to ODE (1) is

    $$T(t) = T_{out} + (T_0 - T_{out})e^{-kt}$$

    where in this context T_{out} is the room temperature. We can measure T_{out} and
    the initial temperature, T_0. Rearranging and taking the natural logarithm of
    both sides gives

    $$\ln|T(t) - T_{out}| = \ln|T_0 - T_{out}| - kt$$

    Using the data of Problem 3, plot $\ln|T(t) - T_{out}|$ against t. What would you
    expect the graph to look like? Use your graph to estimate k, then use ODE
    Architect to check your results.

Answer questions in the space provided, or on attached sheets with carefully labeled graphs. A notepad report using the Architect is OK, too.

Name/Date _____

Course/Section _____

Exploration 3.2. Keeping Your Cool

1. *On again, off again.*
   When a room is cooled by an air conditioner, the unit switches on and off periodically, causing the temperature in the room to oscillate. How does the period of oscillation depend on the following factors?

   - The upper and lower settings of the control temperature
   - The outside temperature
   - The coil temperature, T_{ac}

2. *Keeping your cool for less.*

   The cost of operating an air conditioner depends on how much it runs. Which is the most economical way of cooling a room over a given time period?

   - Set a small difference between the control temperatures, so that the temperature is always close to the average.
   - Allow a large difference between the control temperatures so that the unit switches on and off less frequently.

   Make sure the average of the control temperatures is the same in all your tests.

Answer questions in the space provided, or on
attached sheets with carefully labeled graphs. A
notepad report using the Architect is OK, too.

Name/Date _____

Course/Section _____

Exploration 3.3. The Return of the Melting Snowman

1. *The half-life of a snowman.*

   Use ODE Architect to plot volume vs. time for several different initial snow-man volumes between 5 and 25 ft^3, assuming that $k = 0.1451$ ft/(hr °F) and $T_{out} = 40$°F. For each initial volume Use the Explore feature of ODE Architect to find the time it takes the snowman to melt to half of its original size and make a plot of this "half-life" vs. initial volume. Any conclusions? [To access the Explore feature, click on Solutions on the menu bar and choose Explore. This will bring up a dialog box and a pair of crosshairs in the graphics window. Move the crosshairs to the appropriate point on the solution curve and read the coordinates of that point from the dialog box. Note that the Index entry gives the corresponding line in the Data table.]

2. *Sensitivity to outside temperature.*

   Now fix the snowman's initial volume at 10 ft^3 and use ODE Architect to plot a graph of volume vs. time for several different outside temperatures between 35°F and 45°F, with $k = 0.1451$ ft/(hr °F). Find the time it takes the snowman to melt to 5 ft^3 for each outside temperature used and plot that time against temperature. Describe the shape of the graph.

3. *Other snowmen.*

In developing our snowman model, we assumed that the snowman could be modeled as a sphere. Sometimes snowmen are built by rolling the snow in a way that makes the body cylindrical. How would you model a cylindrical snowman? Which type of snowman melts faster, given the same initial volume and air temperature?

CHAPTER

4 Second-Order Linear Equations

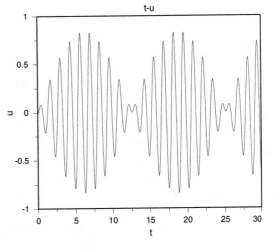

The phenomenon of beats.

Overview Second-order linear differential equations, especially those with constant coefficients, have a host of important applications. In this chapter we explore some
phenomena involving mechanical and electrical oscillations. The first submodule
deals with some basic features common to oscillations of all sorts. The second
submodule applies some of these results to seismographs, which are instruments
used for recording earthquake data.

Key words Oscillation; period; frequency; amplitude; phase; simple harmonic motion; viscous
damping; underdamping; overdamping; critical damping; transient; steady-state
solution; forced oscillation; seismograph; Kirchhoff's laws

See also Chapter 5 for more on vectors and damping, Chapters 6 and 10 for more on oscillations, and Chapter 12 for more on forced oscillations.

◆ Second-Order ODEs and the Architect

ODE Architect will accept only first-order ODEs, so how can we use it to solve a second-order ODE? There is a neat trick that does the job, and an example will show how. Suppose we want to use ODE Architect to study the behavior of the *initial value problem* (or IVP):

$$u'' + 3u' + 10u = 5\cos(2t), \quad u(0) = 1, \quad u'(0) = 0 \qquad (1)$$

Let's write $v = u'$, then

$$v' = \frac{d}{dt}(v) = \frac{d}{dt}(u') = u''$$

but

$$u'' = -10u - 3u' + 5\cos(2t)$$

so IVP (1) becomes

$$\begin{aligned} u' &= v, \quad u(0) = 1 \\ v' &= -10u - 3v + 5\cos(2t), \quad v(0) = 0 \end{aligned} \qquad (2)$$

☞ ODE Architect only accepts ODEs in *normal form*; for example, write $2x' - x = 6$ as $x' = x/2 + 3$ with the x' term alone on the left.

ODE Architect won't accept IVP (1), but it will accept the equivalent IVP (2). The components u and v give the solution of IVP (1) and its first derivative $u' = v$. Therefore, if we use ODE Architect to solve and plot the component curve $u(t)$ of system (2), we are simultaneously plotting the solution $u(t)$ of IVP (1).

✓ "Check" your understanding by converting the IVP

$$2u'' - 2u' + 3u = -\sin(4t), \quad u(0) = -1, \quad u'(0) = 2$$

to an equivalent IVP involving a system of two normalized first-order ODEs.

◆ Undamped Oscillations

Second-order differential equations arise naturally in physical situations; for example, the motion of an object is described by Newton's second law, $F = ma$. Here, a is the acceleration, which is the second derivative of the object's position. Many of these differential equations lead to *oscillations* or *vibrations*. Many oscillating systems can be modeled by a system consisting of a mass attached to a spring where the motion takes place in a horizontal direction on a table. This simplifies the derivation of the equation of motion, but the same equation also describes the up-and-down motion of a mass suspended by a vertical spring.

Let's assume an ideal situation: there is no friction between the mass and the table, there is no air resistance, and there is no dissipation of energy in the

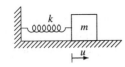

☞ This is also called *Hooke's law restoring force.*

☞ See the first two references for derivations of formula (5).

☞ The term "circular frequency" is only used with trigonometric functions.

☞ This motion is called *simple harmonic motion.* See Screen 1.3 of Module 4 for graphs.

spring or anywhere else in the system. The differential equation describing the motion of the mass is

$$m\frac{d^2u}{dt^2} = -ku \tag{3}$$

where $u(t)$ is the position of the mass m relative to its equilibrium and k is the spring constant. The natural tendency of the spring to return to its equilibrium position is represented by the restoring force $-ku$. Two initial conditions,

$$u(0) = u_0, \qquad u'(0) = v_0 \tag{4}$$

where u_0 and v_0 are the initial position and velocity of the mass, respectively, determine the position of the mass uniquely. ODE (3) together with the initial conditions (4) constitute a well-formulated initial value problem whose solution predicts the position of the mass at any future time.

The general solution of ODE (3) is

$$u(t) = C_1 \cos(\omega_0 t) + C_2 \sin(\omega_0 t) \tag{5}$$

where C_1 and C_2 are arbitrary constants and $\omega_0^2 = k/m$. Applying the initial conditions (4), we find that $C_1 = u_0$ and $C_2 = v_0/\omega_0$. Thus the solution of the IVP (3), (4) is

$$u(t) = u_0 \cos(\omega_0 t) + (v_0/\omega_0) \sin(\omega_0 t) \tag{6}$$

The corresponding motion of the mass is *periodic*, which means that it repeats itself after the passage of a time interval T called the *period*. If we measure time in seconds, then the quantity ω_0 is the *natural (circular) frequency* in radians/sec, and T is given by

$$T = 2\pi/\omega_0 \tag{7}$$

The reciprocal of T, or $\omega_0/2\pi$, is the *frequency* of the oscillations measured in cycles per second, or *hertz*. Notice that since $\omega_0 = \sqrt{k/m}$, the frequency and the period depend only on the mass and the spring constant and not on the initial data u_0 and v_0.

By using a trigonometric identity, the solution (6) can be rewritten in the *amplitude-phase form* as a single cosine term:

$$u(t) = A \cos(\omega_0 t - \delta) \tag{8}$$

where A and δ are expressed in terms of u_0 and v_0/ω_0 by the equations

$$A = \sqrt{u_0^2 + (v_0/\omega_0)^2}, \qquad \tan \delta = \frac{v_0}{u_0 \omega_0} \tag{9}$$

The quantity A determines the magnitude or *amplitude* of the oscillation (8), and δ, called the *phase* (or *phase shift*), measures the time translation from a standard cosine curve.

✓ Show that (8) is equivalent to (6) when A and δ are defined by (9).

◆ The Effect of Damping

☞ The viscous damping force is $-c\,du/dt$.

Equation (8) predicts that the periodic oscillation will continue indefinitely. A more realistic model of an oscillating spring must include damping. A simple, useful model results if we represent the damping force by a single term that is proportional to the velocity of the mass. This model is known as the *viscous damping* model; it leads to the differential equation

$$m\frac{d^2u}{dt^2} + c\frac{du}{dt} + ku = 0 \qquad (10)$$

where the positive constant c is the viscous damping coefficient.

The behavior of the solutions of ODE (10) is determined by the roots r_1 and r_2 of the *characteristic polynomial equation,*

$$mr^2 + cr + k = 0$$

Using the quadratic formula, we find that the *characteristic roots* r_1 and r_2 are

☞ Check that this equation gives a solution of ODE (10).

$$r_1 = \frac{-c + \sqrt{c^2 - 4mk}}{2m}, \quad r_2 = \frac{-c - \sqrt{c^2 - 4mk}}{2m} \qquad (11)$$

The nature of the solutions of ODE (10) depends on the sign of the *discriminant* $c^2 - 4mk$. If $c^2 \neq 4mk$, then $r_1 \neq r_2$ and the general solution of ODE (10) is

$$u = C_1 e^{r_1 t} + C_2 e^{r_2 t} \qquad (12)$$

where C_1 and C_2 are arbitrary constants.

The most important case is *underdamping* and occurs when $c^2 - 4mk < 0$, which means that the damping is relatively small. In the underdamped case, the characteristic roots r_1 and r_2 in formula (11) are the complex numbers

$$r_1 = -\frac{c}{2m} + i\mu, \quad r_2 = -\frac{c}{2m} - i\mu, \quad \text{where } \mu = \frac{\sqrt{4mk - c^2}}{2m} \neq 0 \quad (13)$$

Euler's formula implies that

$$e^{(\alpha + i\beta)t} = e^{\alpha t}(\cos \beta t + i \sin \beta t)$$

for any real numbers α and β, so

$$e^{r_1 t} = e^{-ct/2m}(\cos \mu t + i \sin \mu t), \quad e^{r_2 t} = e^{-ct/2m}(\cos \mu t - i \sin \mu t) \quad (14)$$

Now, using the initial conditions together with equations (12) and (14), we find that the solution of the IVP

$$m\frac{d^2u}{dt^2} + c\frac{du}{dt} + ku = 0, \quad u(0) = u_0, \quad u'(0) = v_0 \qquad (15)$$

☞ Solutions of an underdamped ODE oscillate with circular frequency μ and an exponentially decaying amplitude.

is given by

$$u = e^{-ct/2m}\left\{ u_0 \cos(\mu t) + \left[\frac{v_0}{\mu} + \frac{cu_0}{2m\mu} \right] \sin(\mu t) \right\} \qquad (16)$$

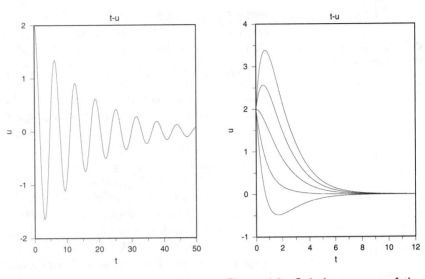

Figure 4.1: A solution curve of the underdamped spring-mass ODE, $u'' + 0.125u' + u = 0$.

Figure 4.2: Solution curves of the overdamped spring-mass ODE, $u'' + 2.1u' + u = 0$.

✓   Verify that $u(t)$ defined in formula (16) is a solution of IVP (15).

The solution (16) represents an oscillation with circular frequency μ and an exponentially decaying amplitude (see Figure 4.1). From the formula in (13) we see that $\mu < \omega_0$, where $\omega_0 = \sqrt{k/m}$, but the difference is small for small c.

☞   Take a look at Screen 1.6 of Module 4.

If the damping is large enough so that $c^2 - 4mk > 0$, then we have *over-damping* and the solution of IVP (15) decays exponentially to the equilibrium position but does not oscillate (see Figure 4.2). The transition from oscillatory to nonoscillatory motion occurs when $c^2 - 4mk = 0$. The corresponding value of c, given by $c_0 = 2\sqrt{mk}$, is called *critical damping*.

◆ Forced Oscillations

☞ $F(t)$ is also called the *input*, or *driving term*; solutions $u(t)$ are the *responses* to the input and the initial data.

Now let's see what happens when an external force is applied to the oscillating mass described by ODE (10). If $F(t)$ represents the external force, then ODE (10) becomes

$$m\frac{d^2u}{dt^2} + c\frac{du}{dt} + ku = F(t) \tag{17}$$

Some interesting things happen if $F(t)$ is periodic, so we will look at the ODE

$$m\frac{d^2u}{dt^2} + c\frac{du}{dt} + ku = F_0\cos(\omega t) \tag{18}$$

☞ Check that this formula gives solutions of ODE (18).

where F_0 and ω are the amplitude and circular frequency, respectively, of the external force F. Then, in the underdamped case, the general solution of ODE (18) has the form

$$u(t) = e^{-ct/2m}[C_1 \cos(\mu t) + C_2 \sin(\mu t)] + a \cos(\omega t) + b \sin(\omega t) \qquad (19)$$

where a and b are constants determined so that $a \cos(\omega t) + b \sin(\omega t)$ is a solution of ODE (18). The constants a and b depend on m, c, k, F_0, and ω of ODE (18), but not on the initial data. The constants C_1 and C_2 can be chosen so that $u(t)$ given by formula (19) satisfies given initial conditions.

The first term on the right side of the solution (19) approaches zero as $t \to +\infty$; this is called the *transient* term. The remaining two terms do not diminish as t increases, and their sum is called the *steady-state solution* (or the *forced oscillation*), here denoted by $u_s(t)$. Since the steady-state solution persists forever with constant amplitude, it is frequently the most interesting solution. Notice that it oscillates with the circular frequency ω of the driving force F. It can be written in the amplitude-phase form (8) as

$$u_s(t) = A \cos(\omega t - \delta) \qquad (20)$$

where A and δ are now given by

$$A = \frac{F_0}{\sqrt{m^2(\omega_0^2 - \omega^2)^2 + c^2\omega^2}}, \quad \tan\delta = \frac{c\omega}{m(\omega_0^2 - \omega^2)} \qquad (21)$$

Figure 4.3 shows a graphical example of solutions that tend to a forced oscillation.

For an underdamped system with fixed c, k, and m, the amplitude A of the steady-state solution depends upon the frequency of the driving force. It

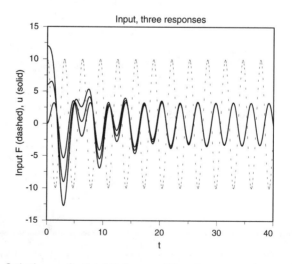

Figure 4.3: Solutions of $u'' + 0.3u' + u = 10\cos 2t$ approach a unique forced oscillation with the circular frequency 2 of the input.

is important to know whether there is a value $\omega = \omega_r$ for which the amplitude is maximized. If so, then driving the system at the circular frequency ω_r produces the greatest response. Using methods of calculus, it can be shown that if $c^2 < 2mk$ then ω_r is given by

☞ Recall that the *natural circular frequency* ω_0 is the value $\omega_0 = \sqrt{k/m}$.

$$\omega_r^2 = \omega_0^2 \left(1 - \frac{c^2}{2mk}\right) \tag{22}$$

The corresponding maximum value A_r of the amplitude when $\omega = \omega_r$ is

$$A_r = \frac{F_0}{c\omega_0\sqrt{1 - (c^2/4mk)}} \tag{23}$$

✓ Does A have a maximum value when $2mk < c^2 < 4mk$?

✓ Find the forced oscillation for the ODE of Figure 4.3.

◆ Beats

Let's polish the table and streamline the mass so that damping is negligible. Then we apply a forcing function whose frequency is close to the natural frequency of the spring-mass system, and watch the response. We can model this by the IVP

$$u'' + \omega_0^2 u = \frac{F_0}{m}\cos(\omega t), \quad u(0) = 0, \quad u'(0) = 0 \tag{24}$$

where $|\omega_0 - \omega|$ is small (but not zero). The solution is

$$u(t) = \frac{F_0}{m(\omega_0^2 - \omega^2)}[\cos(\omega t) - \cos(\omega_0 t)]$$

$$= \left[\frac{2F_0}{m(\omega_0^2 - \omega^2)}\sin\left(\frac{\omega_0 - \omega}{2}t\right)\right]\sin\left(\frac{\omega_0 + \omega}{2}t\right) \tag{25}$$

where trigonometric identities have been used to get from the first form of the solution to the second. The term in square brackets in formula (25) can be viewed as a varying amplitude for the sinusoid term $\sin[(\omega_0 + \omega)/2]t$. Since $|\omega_0 - \omega|$ is small, the circular frequency $(\omega_0 + \omega)/2$ is much higher than the low circular frequency $(\omega_0 - \omega)/2$ of the varying amplitude. Therefore we have a rapid oscillation with a slowly varying amplitude. This is the *beat phenomenon* illustrated on the chapter cover figure for the IVP

$$u'' + 25u = 2\cos(4.5t), \quad u(0) = 0, \quad u'(0) = 0$$

If you try this out with a driven mass on a spring you will see rapid oscillations whose amplitude slowly grows and then diminishes in a repeating pattern. This phenomenon can actually be heard when a pair of tuning forks which have nearly equal frequencies are struck simultaneously. We hear the "beats" as each acts as a driving force for the other.

◆ Electrical Oscillations: An Analogy

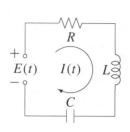

Linear differential equations with constant coefficients are important because they arise in so many different physical contexts. For example, an ODE similar to ODE (17) can be used to model charge oscillations in an electrical circuit. Suppose an electrical circuit contains a resistor, an inductor, and a capacitor connected in series. The current I in the circuit and the charge Q on the capacitor are functions of time t. Let's assume we know the resistance R, the inductance L, and the capacitance C. By Kirchhoff's voltage law for a closed circuit, the applied voltage $E(t)$ is equal to the sum of the voltage drops through the various elements of the circuit. Observations of circuits suggest that these voltage drops are as follows:

- The voltage drop through the resistor is RI (Ohm's law);
- The voltage drop through the inductor is $L(dI/dt)$ (Faraday's law);
- The voltage drop through the capacitor is Q/C (Coulomb's law).

Thus, by Kirchhoff's law, we obtain the differential equation

$$L\frac{dI}{dt} + RI + \frac{Q}{C} = E(t) \tag{26}$$

Since $I = dQ/dt$, we can write ODE (26) entirely in terms of Q,

$$L\frac{d^2Q}{dt^2} + R\frac{dQ}{dt} + \frac{Q}{C} = E(t) \tag{27}$$

ODE (27) models the charge $Q(t)$ on the capacitor of what is called the *simple RLC circuit* with voltage source $E(t)$. ODE (27) is equivalent to ODE (17), except for the symbols and their interpretations. Therefore we can also apply conclusions about our spring-mass system to electrical circuits. For example, we can interpret the ODE $u'' + 0.3u + u = 10\cos 2t$, whose solutions are graphed in Figure 4.3, as a model either for the oscillations of a damped and driven spring-mass system, or the charge on the capacitor of a driven *RLC* circuit. We see that a mathematical model can have many interpretations, and any mathematical conclusions about the model apply to every interpretation.

✓   What substitutions of parameters and variables would you have to make in ODE (27) to transform it to ODE (17)?

◆ Seismographs

☞   Look at "Earthquakes and the Richter Scale" in Module 4.

Seismographs are instruments that record the displacement of the ground as a function of time, and a seismometer is the part of a seismograph that responds to the motion. Seismographs come in two generic types. Matt's friend Seismo is a horizontal-component seismograph, which records one of the horizontal components of the earth's local motion. Of course, two horizontal components are required to specify fully horizontal motion, usually by means of north-south and east-west components. The other type of seismograph records the vertical component of motion. Both of these instruments are based on pendulums that respond to the motion of the ground relative to the seismograph.

Since Seismo is an animation of a horizontal-component seismograph, we'll outline the derivation of the ODEs that govern the motion of his arm. The starting point is the angular form of Newton's second law of motion, also known as the *angular momentum law*:

☞ If you're queasy about cross products or approximating functions (as we do in formula (29)) you may prefer to skip directly to ODE (33) or ODE (34).

$$\frac{d}{dt}\mathbf{L} = \mathbf{R} \times \mathbf{F} \tag{28}$$

where $\mathbf{L}$ is the angular momentum of a mass (Seismo's arm and hand) about a fixed axis, $\mathbf{F}$ is the force acting on the mass, $\mathbf{R}$ is the position vector from the center of mass of Seismo's arm and hand to the axis, and $\times$ is the vector cross product.

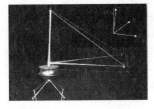

We'll apply this law using an orthogonal xyz-coordinate system which is illustrated on Screen 2.2 of Module 4. In this system the y-axis is horizontal. Seismo's body is parallel to the z-axis and the rest position of Seismo's arm is parallel to the x-axis. The z-axis is not parallel to the local vertical, but instead is the axis which results from rotating the local vertical through a small angle α about the y-axis. Because of this small tilt, the x-axis points slightly downward and the arm is in a stable equilibrium position when it is parallel to the x-axis. The seismic disturbance is assumed to be in the direction of the y-axis. The xz-plane is called Seismo's *rest plane*.

Seismo's hand writes on the paper in the xy-plane, and the angle θ measures the angular displacement of his arm from its rest position. Consider an axis pointing in the z-direction and through the center of mass of Seismo's arm and hand, and let m represent the mass of the arm and hand. The z-component of the angular momentum about that axis is $mr^2(d\theta/dt)$ where r is the radius of gyration of the arm.

To compute the right-hand side of ODE (28), we need to know $\mathbf{R}$, the position vector from the center of mass of Seismo's arm and hand to the origin. Note that

$$\mathbf{R} = -l\cos\theta\hat{\mathbf{x}} - l\sin\theta\hat{\mathbf{y}}$$

where l is the distance from the center of mass to Seismo's body, and $\hat{\mathbf{x}}$ and $\hat{\mathbf{y}}$ are unit vectors along the positive x- and y-axes. For small θ, we have the approximations $\cos\theta \approx 1$ and $\sin\theta \approx \theta$, so

$$\mathbf{R} = -l\hat{\mathbf{x}} - l\theta\hat{\mathbf{y}} \tag{29}$$

Using equation (29) in ODE (28) and computing the cross product, we obtain

$$mr^2\frac{d^2\theta}{dt^2} = -lF^{(y)} + l\theta F^{(x)} \tag{30}$$

indicating by superscripts the components of the net force $\mathbf{F}$ exerted on the arm and hand.

Now we need expressions for the two components of $\mathbf{F}$ in ODE (30). If the x-component of friction is assumed negligible then the two force components acting in the x-direction are the x-component of the gravitational force and the x-component of the force due to the seismic disturbance. Because the arm displacement angle θ and the body inclination angle α are both assumed small, the x-component of the force due to the seismic disturbance can be shown to be negligible also. Therefore the x-component of the net force, $F^{(x)}$, is given by the simple form

$$F^{(x)} \approx mg\alpha \qquad (31)$$

The right side of equation (31) is the gravity component $mg\sin\alpha$ approximated by $mg\alpha$.

☞ This is *viscous friction*.

In the y-direction, the forces acting are the force due to the seismic disturbance and to friction, the latter assumed to be proportional to the angular velocity $d\theta/dt$. The force due to the seismic disturbance can be computed as follows: Let h be a small ground displacement in the y-direction. Then the y-coordinate of the center of mass is approximately $h + l\theta$. Therefore the force due to the earthquake is approximated by

$$m\frac{d^2}{dt^2}(h + l\theta) = m\frac{d^2h}{dt^2} + ml\frac{d^2\theta}{dt^2}$$

and the net force in the y-direction is

$$F^{(y)} \approx m\frac{d^2h}{dt^2} + ml\frac{d^2\theta}{dt^2} - k\frac{d\theta}{dt} \qquad (32)$$

where k is a positive constant characterizing the effect of friction.

Combining ODE (30) with formulas (31) and (32), we find that the motions of Seismo's arm are governed by the ODE

$$\frac{d^2\theta}{dt^2} + c\frac{d\theta}{dt} + \omega_0^2\theta = -\frac{1}{L}\frac{d^2h}{dt^2} \qquad (33)$$

In ODE (33) the quantities ω_0^2, L, and c are given by $\omega_0^2 = g\alpha/L$, where $L = (r^2 + l^2)/l$, and $c = k/(mL)$. We can interpret the terms in (33) as follows. The first term on the left arises from the inertia of Seismo's hand and arm. The second term models the frictional force due to the angular motion of the arm. The third term, arising from gravity and the tilt of the arm, is the restoring force for the oscillations of the arm and hand. Finally, the term on the right arises from the effective force of the seismic displacement.

To simplify ODE (33) a little more, we let $h(t) = Hf(t)$, where H is the maximum ground displacement, which means that the maximum value of the dimensionless ground displacement $f(t)$ is one. Then ODE (33) becomes the following equation for the dimensionless arm displacement $y(t) = L\theta(t)/H$:

$$\frac{d^2y}{dt^2} + c\frac{dy}{dt} + \omega_0^2 y = -\frac{d^2f}{dt^2} \qquad (34)$$

This is the ODE used in Screen 2.3 of Module 4.

☞ A given ODE can model a variety of phenomena.

It's an important result that ODE (34) is the same type as ODE (18), the only differences being in the definitions of the parameters that multiply the individual terms, and in the choice of variables. A second striking result is that this same ODE (34) applies as well to the motions of a vertical component seismograph. All that is necessary are other modifications in the meanings of the parameters and functions. Details are available in the book by Bullen and Bolt in the references.

References

Borrelli, R. L., and Coleman, C. S., *Differential Equations: A Modeling Perspective*, (1998: John Wiley & Sons, Inc.)

Boyce, W. E., and DiPrima, R. C., *Elementary Differential Equations and Boundary Value Problems*, 6th ed. (1997: John Wiley & Sons, Inc.)

Bullen, K. E., and Bolt, B. A., *An Introduction to the Theory of Seismology*, 4th ed. (1985: Cambridge University Press)

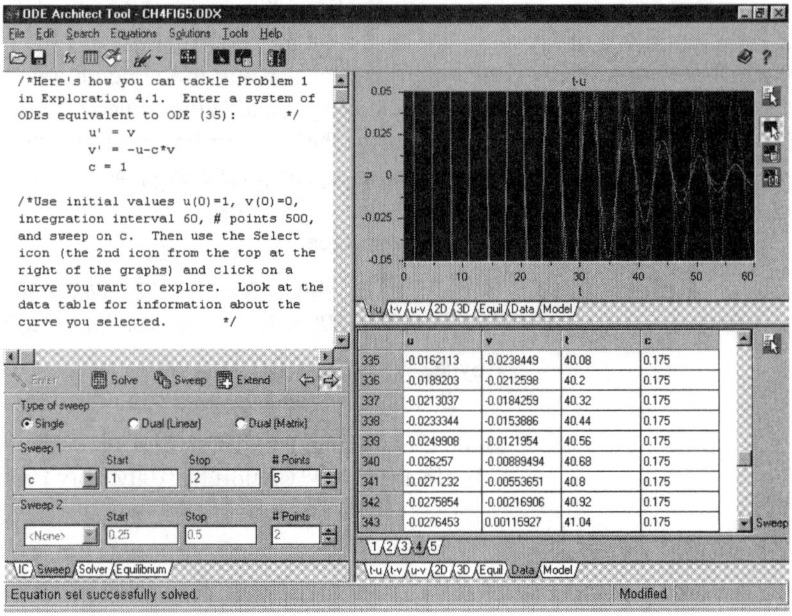

Figure 4.4: The sweep on c generated five solution curves. The selected curve is highlighted, and the corresponding solution $u(t)$ satisfies the condition $|u(t)| < 0.05$ for $t \geq 40$. The data tells us that $c = 0.175$ for the curve.

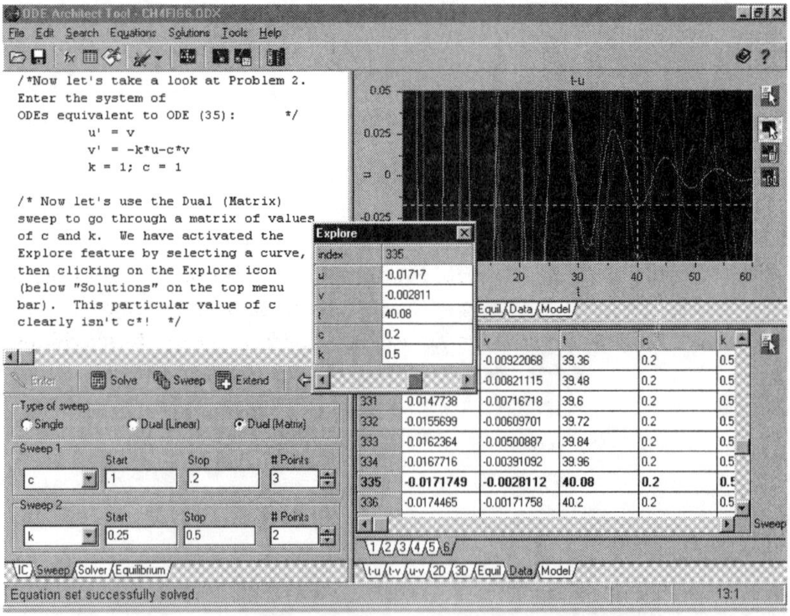

Figure 4.5: The Dual (Matrix) feature produces six solutions for various values of c and k. We have selected one of them (the highlighted curve) and used the Explore option to get additional information.

Answer questions in the space provided, or on
attached sheets with carefully labeled graphs. A
notepad report using the Architect is OK, too.

Name/Date _____

Course/Section _____

Exploration 4.1. The Damping Coefficient

Assume that Dogmatic's oscillations satisfy the IVP

$$u'' + cu' + ku = 0, \quad u(0) = 1, \quad u'(0) = 0 \tag{35}$$

1. Let $k = 1$ and use ODE Architect to estimate the smallest value c^* of the damping coefficient c so that $|u(t)| \leq 0.05$ for all $t \geq 40$. [*Suggestion:* Figure 4.4 illustrates one way to estimate c^* by using the Select feature and the Data table.]

2. Repeat Problem 1 for other values of k, including $k = \frac{1}{4}, \frac{1}{2}, 2$, and 4. How does c^* change as k changes? [*Suggestion:* Figure 4.5 shows the outcome of using a Dual (Matrix) sweep on the values of c and k, and then using the Explore feature.]

3. Let $k = 10$ in IVP (35).

(a) Find the value of c for which the ratio of successive maxima in the graph of u vs. t is 0.75.

(b) Why is the ratio between successive maxima always the same?
Note: Since the values of the maxima can be observed experimentally, this provides a practical way to determine the value of the damping coefficient c, which may be difficult to measure directly.

Answer questions in the space provided, or on
attached sheets with carefully labeled graphs. A
notepad report using the Architect is OK, too.

Name/Date _____

Course/Section _____

Exploration 4.2. Response to the Forcing Frequency

1. Suppose that Dogmatic's oscillations satisfy the differential equation

$$2u'' + u' + 4u = 2\cos(\omega t)$$

   Let $\omega = 1$. Select your own initial conditions and use ODE Architect to plot the solution over a long enough time interval that the transient part of the solution becomes negligible. From the graph, determine the amplitude A_s of Dogmatic's steady-state solution.

2. Repeat Problem 1 for other values of ω. Plot the corresponding pairs ω, A_s and sketch the graph of A_s vs. ω. Estimate the value of ω for which A_s is a maximum. *Note*: You may want to use the Lookup Table feature of ODE Architect (see Module 1 and Chapter 1 for details).

3. In Problems 1 and 2, the value of the damping coefficient c is 1. Repeat your calculations for $c = \frac{1}{2}$ and $c = \frac{1}{4}$. How does the maximum value of A_s change as the value of c changes? Compare your results with the predictions of formula (23).

Answer questions in the space provided, or on
attached sheets with carefully labeled graphs. A
notepad report using the Architect is OK, too.

Name/Date _____

Course/Section _____

Exploration 4.3. Low- and High-Frequency Quakes

In experiments with Seismo, you used ODE (34) to find the response of his arm to different ground displacements of sinusoidal type, $f(t) = \cos \omega t$, when $1 \leq \omega \leq 5$. In this exploration you'll investigate what happens for ground displacements with frequencies that are lower or higher than these values.

1. Choose $c = 2$ and $\omega_0 = 3$ in ODE (34), and set the initial conditions $y(0)$ and $y'(0)$ to zero. Use $f(t) = \cos \omega t$ with $\omega = 0.5$ for the ground displacement. Use ODE Architect to plot the displacement $y(t)$ determined from ODE (34); also plot $f(t)$ on the same graph. How do the features of $y(t)$ compare with those of $f(t)$?

2. Repeat Problem 1 for values of ω smaller than 0.5. Be sure to plot for a long enough time interval to see the relevant time variations. What do you think is Seismo's arm response as ω approaches zero? How does this compare with the corresponding response of a mass on a spring from ODE (18)?

3. Repeat Problem 1 for values of ω larger than 5, such as $\omega = 10$ and $\omega = 20$. What do you think is Seismo's arm response as ω becomes very large?

Answer questions in the space provided, or on attached sheets with carefully labeled graphs. A notepad report using the Architect is OK, too.

Name/Date _____

Course/Section _____

Exploration 4.4. Different Ground Displacements

In explorations with Seismo, we assumed that the dimensionless ground displacements $f(t)$ are sinusoidal, with a single frequency. Real earthquakes however, are not so simple: you'll investigate other possibilities in the following problems. The ODE for Seismo's dimensionless arm displacement $y(t)$ is

$$\frac{d^2y}{dt^2} + c\frac{dy}{dt} + \omega_0^2 y = -\frac{d^2f}{dt^2} \tag{36}$$

1. Suppose the ground displacement can be modeled by the function

$$f(t) = \begin{cases} (t/T)^2, & 0 \le t \le T \\ 1, & t > T \end{cases}$$

How do you interpret this motion? Choose $c = 2$ and $\omega_0 = 3$, and set $y(0) = y'(0) = 0$. Use ODE Architect to find $y(t)$ from ODE (36) for the case $T = 2$, and display both $y(t)$ and $f(t)$. Note: d^2f/dt^2 can be written using a step function. How do the features of $y(t)$ compare with those of $f(t)$? For example, what is the maximum magnitude of $y(t)$, and when does it occur?

2. Now suppose that the ground motion is given by the function $f = e^{-at} \sin(\pi t)$. Choose some values of a in the range $0 < a \leq 0.5$ and study how Seismo's arm displacements change with the parameter a.

3. How do you think the results of Problem 2 would change if the period of the sinusoidal oscillation were different from 2? Try a few cases to check your predictions.

5 Models of Motion

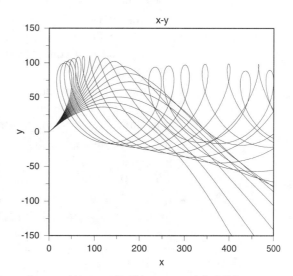

Seventeen ski jumpers take off from an upward-tilted ski-jump.

Overview   How would you model the motion of a baseball thrown at a target, or the rise and fall of a whiffle ball, or the trajectory of a ski jumper? You need modeling principles to explain the effects of the surroundings on the motion of a body.

Building on the work of Galileo, Newton formulated the fundamental laws of motion that describe the forces acting on a body in terms of the body's acceleration and mass. Newton's second law of motion, for example, relates the mass and the acceleration of a moving body to the forces acting on it and ultimately leads to differential equations for the motion.

Bodies moving through the air near the surface of the earth (e.g., a whiffle ball, Indiana Newton jumping onto a boxcar, or a ski jumper) are subject to the forces of gravity and air resistance, so these forces will affect their motion.

Key words   Vectors; force; gravity; Newton's laws; acceleration; trajectory; air resistance; viscous drag; Newtonian drag; lift

See also   Chapter 1 for more on modeling, and Chapter 2 for "The Juggler" and "The Sky Diver".

◆ Vectors

A *vector* is a directed line segment and can be represented by an arrow with a head and a tail. We use boldface letters to denote vectors.

Some terminology:

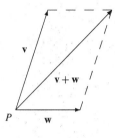

- The *length* of a vector **v** is denoted by $|\mathbf{v}|$.

- Two vectors **v** and **w** are *equivalent* if they can be made to coincide by translations. (Translations preserve length and direction of vectors.) So parallel vectors of equal length and pointing in the same direction are equivalent.

- The *sum* $\mathbf{v} + \mathbf{w}$ of **v** and **w** is defined by the parallelogram law as follows: $\mathbf{v} + \mathbf{w}$ is the diagonal vector of the parallelogram formed by **v** and **w** as shown in the margin figure.

- If r is any real number, then the *product* $r\mathbf{v}$ is the vector of length $|r|\,|\mathbf{v}|$ that points in the direction of **v** if $r > 0$ and in the direction opposite to **v** if $r < 0$.

- If a vector $\mathbf{u} = \mathbf{u}(t)$ depends on a variable t, then the *derivative* $d\mathbf{u}/dt$ [or $\mathbf{u}'(t)$] is defined as the limit of a difference quotient:

$$\mathbf{u}'(t) = \frac{d\mathbf{u}}{dt} = \lim_{h \to 0} \frac{\mathbf{u}(t+h) - \mathbf{u}(t)}{h}$$

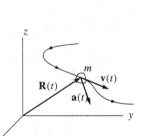

- A *coordinate frame* is a triple of vectors, denoted by $\{\hat{\mathbf{i}}, \hat{\mathbf{j}}, \hat{\mathbf{k}}\}$, that are mutually orthogonal and all of unit length. Every vector can be uniquely written as the sum of vectors parallel to $\hat{\mathbf{i}}$, $\hat{\mathbf{j}}$, and $\hat{\mathbf{k}}$. So for each vector **v** there is a unique set of real numbers v_1, v_2, and v_3 such that $\mathbf{v} = v_1\hat{\mathbf{i}} + v_2\hat{\mathbf{j}} + v_3\hat{\mathbf{k}}$. Here v_1, v_2, and v_3 are called the *coordinates* (or *components*) of **v** in the frame $\{\hat{\mathbf{i}}, \hat{\mathbf{j}}, \hat{\mathbf{k}}\}$.

Let's see how to use vectors in a real-life situation. Suppose a particle of mass m moves in a manner described by the *position vector*

$$\mathbf{R} = \mathbf{R}(t) = x(t)\hat{\mathbf{i}} + y(t)\hat{\mathbf{j}} + z(t)\hat{\mathbf{k}}$$

If **R** is differentiable, then

$$\mathbf{R}'(t) = x'(t)\hat{\mathbf{i}} + y'(t)\hat{\mathbf{j}} + z'(t)\hat{\mathbf{k}}$$

The vector $\mathbf{R}'(t) = \mathbf{v}(t)$ is the *velocity vector* of the particle at time t, and $\mathbf{v}(t)$ is tangent to the path of the particle's motion at the point $\mathbf{R}(t)$. If $\mathbf{R}'(t)$ is differentiable, then

$$\mathbf{R}''(t) = \mathbf{v}'(t) = x''(t)\hat{\mathbf{i}} + y''(t)\hat{\mathbf{j}} + z''(t)\hat{\mathbf{k}}$$

The vector $\mathbf{R}''(t) = \mathbf{a}(t)$ is the *acceleration vector* for the particle.

✓   "Check" your understanding by answering this question: If a particle moves at a constant speed around a circle, does the acceleration vector from the particle point to the inside of the circle or to the outside of the circle?

✓   If a particle's acceleration vector is always tangent to its path, what is the path?

Next, let's use vectors to express Newton's laws of motion.

◆ Forces and Newton's Laws

☞ Deceleration is just negative acceleration.

Our environment creates forces that act on bodies in a way that causes the bodies to accelerate or decelerate. Forces have magnitudes and directions and so can be represented by vectors. Newton formulated two laws that describe how the forces on a body relate to its motion.

> **Newton's First Law.** A body remains in a state of rest, or in a state of uniform motion in a straight line if there is no net external force acting on it.

But the more interesting situation is when there *is* a net external force acting on the body.

> **Newton's Second Law.** For a body with acceleration **a** and constant mass m,
>
> $$\mathbf{F} = m\mathbf{a}$$
>
> where **F** is the sum of all external forces acting on the body.

Sometimes it's easier to visualize Newton's second law in terms of the x-, y-, and z-components of the position vector **R** of the moving body. If we project the acceleration vector $\mathbf{a} = \mathbf{R}''$ and the forces onto the x-, y-, and z-axes, then for a body of mass m,

$$mx'' = \text{the sum of the forces in the } x\text{-direction}$$
$$my'' = \text{the sum of the forces in the } y\text{-direction}$$
$$mz'' = \text{the sum of the forces in the } z\text{-direction}$$

We'll look at motion in a plane with x measuring the horizontal distance and y measuring the vertical distance up from the ground. We don't need the z-axis for our examples because the motion is entirely along a line or in a plane.

◆ Dunk Tank

Imagine your favorite professor seated over a dunk tank. Let's construct a model that will help you find the secret to hitting the target and giving your teacher a swim!

You hurl a ball at the target from a height of 6 ft with speed v_0 ft/sec and with a launch angle of θ_0 radians from the horizontal[1]. The target is centered 10 ft above the ground and 20 ft away. Let's suppose that air resistance doesn't have much effect on the ball over its short path, so that gravity, acting downwards, is the only force acting on the ball.

Newton's second law says that

$$m\mathbf{R}'' = -mg\hat{\mathbf{j}}$$

where m is the ball's mass, $\mathbf{R}(t)$ is the position of the ball at time t relative to your hand (which is 6 ft above the ground at the instant $t = 0$ of launch), and $g = 32$ ft/sec^2 is the acceleration due to gravity. In coordinate terms,

$$mx'' = 0$$
$$my'' = -mg$$

Since $x'(0) = v_0 \cos\theta_0$ and $y'(0) = v_0 \sin\theta_0$, one integration of these second-order ODEs gives us

$$x'(t) = v_0 \cos\theta_0$$
$$y'(t) = v_0 \sin\theta_0 - gt \tag{1}$$

☞ What is the ball doing if $\theta_0 = \pi/2$?

Then because $x(0) = 0$ and $y(0) = 6$, a second integration yields

$$x(t) = (v_0 \cos\theta_0)t$$
$$y(t) = 6 + (v_0 \sin\theta_0)t - \frac{1}{2}gt^2 \tag{2}$$

To hit the target at some time T we want $x(T) = 20$ and $y(T) = 10$. So values of $T > 0$, θ_0, and v_0 such that

$$x(T) = 20 = (v_0 \cos\theta_0)T$$
$$y(T) = 10 = 6 + (v_0 \sin\theta_0)T - \frac{1}{2}gT^2 \tag{3}$$

lead to hitting the target right in the bull's eye and dunking your professor.

You can try to use system (3), or you can just adjust your launch angle and pitching speed by intuition and experience. The screen shot in Figure 5.1 shows you how to get started with the latter approach. If you play the dunking game on Screen 1.3 you'll find that you can dunk without hitting the target head-on, but that a little up or a little down from the center works fine.

---

[1] The sin, cos and other trig functions in the ODE Architect Tool expect angles to be measured in radians. Note that $\theta_0 = 1$ radian corresponds to $360/2\pi \approx 57.3$ degrees. The multimedia modules will accept angles measured in degrees.

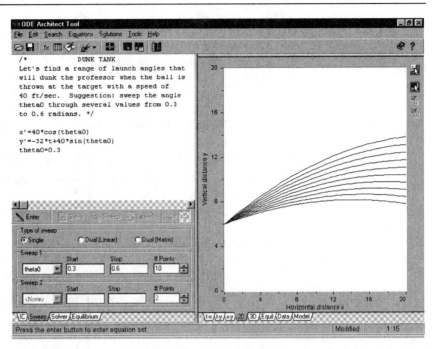

Figure 5.1: This ODE Architect screen shows paths of a ball thrown at ten different angles θ_0. Which paths lead to dunking the prof?

◆ Longer to Rise or to Fall?

Throw a ball straight up in the air and ask observers whether the ball takes longer to rise or to fall. You'll get four answers:

1. Longer to rise
2. Longer to fall
3. Rise-time and fall-time are the same
4. It all depends . . .

What's your answer?

A mathematical model and ODE Architect suggest the answer. The forces acting on the ball of mass m are gravity and air resistance, so Newton's second law states that

$$m\mathbf{R}''(t) = -mg\hat{\mathbf{j}} + \mathbf{F}$$

where $\mathbf{R}$ is the position vector of the ball, and $\mathbf{F}$ is the drag on the ball caused by air resistance. In this case $\mathbf{R}(t) = y(t)\hat{\mathbf{j}}$ where $\hat{\mathbf{j}}$ is a unit vector pointing upward (the positive y direction). If the drag is negligible, we can set $\mathbf{F} = 0$. For a light ball with an extended surface, like a whiffle ball, the drag, called *viscous drag*, exerts a force approximately proportional to the ball's velocity but opposite in direction:

$$\mathbf{F}(\mathbf{v}) = -k\mathbf{v} = -ky'\hat{\mathbf{j}}$$

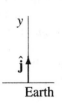

☞ Drag forces are usually determined by observation. They differ widely from one body to another.

If the ball is solid and dense, like a baseball or a bowling ball, then we have *Newtonian drag*, which acts opposite to the velocity with magnitude proportional to the square of the speed:

$$\mathbf{F}(\mathbf{v}) = -k|\mathbf{v}|\mathbf{v} = -k|y'|y'\hat{\mathbf{j}}$$

Summarizing, we have the models

$$y'' = -g - \begin{cases} 0 & \text{no drag} \\ \dfrac{k}{m}y' & \text{viscous} \\ \dfrac{k}{m}|y'|y' & \text{Newtonian} \end{cases}$$

or, in system form,

$$y' = v$$

☞ ODE Architect only accepts first-order ODEs, so that's why we use the first-order system form.

$$v' = -g - \begin{cases} 0 & \text{no drag} \\ \dfrac{k}{m}v & \text{viscous} \\ \dfrac{k}{m}|v|v & \text{Newtonian} \end{cases}$$

To observe different rise times and fall times, you can set $y(0) = 0$, $v(0) = v_0$ and see what happens for various positive values of v_0. See Figure 5.2 for graphs of $y(t)$ with viscous damping, four different initial velocities, $k/m = 2$ sec^{-1}, and $g = 32$ ft/sec^2. In this setting v is the rate of change of y, so v is positive as the ball rises and negative as it falls.

◆ Indiana Newton

You notice that Indiana Newton is about to jump from a ledge onto a boxcar of a speeding train. His timing has to be perfect. He also gets to choose his drag: none, viscous, or Newtonian. If you knew the train's position at all times, and how long it takes Indy to drop from the ledge to the top of the boxcar, then you could give him good advice about which drag to choose.

The initial value problem that models Indy's situation is

$$\begin{aligned} y' &= v & y(0) &= h \\ v' &= -g - F(v)/m & v(0) &= 0 \end{aligned}$$

where m is his mass, $F(v)$ is a drag function, g is the acceleration due to gravity, and h is the height of the ledge above the boxcar. His life is in your hands! Figure 5.3 shows Indy's free-fall solution curves $y(t)$ from a height of 100 ft with three different drag functions.

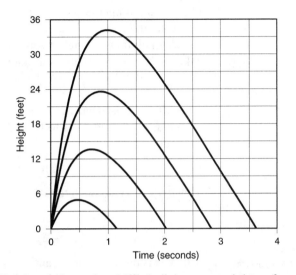

Figure 5.2: Height vs. time of a whiffle ball thrown straight up four times with viscous damping and different initial velocities. Does the ball take longer to rise or to fall?

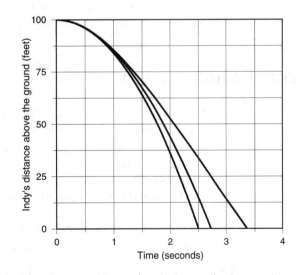

Figure 5.3: Indiana jumps with no drag (left curve), viscous drag $-0.2y'$ (middle), Newtonian drag $-0.02|y'|y'$ (right).

◆ Ski Jumping

When a ski jumper is aloft she is subject to gravitational, drag, and lift forces. She can diminish the drag and increase the lift by her posture, ski angle, and choice of clothing. *Drag* acts opposite to velocity and its magnitude is usually taken to be proportional to the skier's velocity $\mathbf{R}'$:

$$\text{Drag force} = -\delta\mathbf{R}' = -\delta x'\hat{\mathbf{i}} - \delta y'\hat{\mathbf{j}}$$

The lift force is what makes ski jumping fun. The *lift force* is that force which acts perpendicular to the velocity and enables the jumper to soar. Its magnitude is usually taken to be proportional to the speed, so

☞ Check that this force is perpendicular to velocity.

$$\text{Lift force} = -\lambda y'\hat{\mathbf{i}} + \lambda x'\hat{\mathbf{j}}$$

Newton's second law in the $\hat{\mathbf{i}}$- and $\hat{\mathbf{j}}$-directions gives us

☞ The origin of the xy-plane is at the edge of the ski jump (x-horizontal, y-vertical). The edge is horizontal so $x'(0) = v_0 > 0$, but $y'(0) = 0$.

$$mx'' = -\delta x' - \lambda y' \qquad x'(0) = v_0, \qquad x(0) = 0$$
$$my'' = -mg + \lambda x' - \delta y' \qquad y'(0) = 0, \qquad y(0) = 0$$

where m is the skier's mass and δ, λ, and v_0 are positive constants. Integration of each of these ODEs yields

$$mx' - mv_0 = -\delta x - \lambda y$$
$$my' = -mgt + \lambda x - \delta y$$

Divide by the mass to get the system IVP

$$x' = -ax - by + v_0 \qquad x(0) = 0$$
$$y' = -gt + bx - ay \qquad y(0) = 0$$

where $a = \delta/m$ and $b = \lambda/m$ are the drag and lift coefficients, respectively.

When Newtonian drag and lift occur, δ and λ are not constants, so we can no longer integrate once to get x' and y', and we must treat the original second-order ODE differently:

$$x' = v \qquad\qquad x(0) = 0$$
$$v' = -\delta v/m - \lambda w/m \qquad v(0) = v_0$$
$$y' = w \qquad\qquad y(0) = 0$$
$$w' = -g + \lambda v/m - \delta w/m \qquad w(0) = 0$$

where v and w are the velocities in the $\hat{\mathbf{i}}$- and $\hat{\mathbf{j}}$-directions, respectively.

We have assumed that the bottom edge of the ski jump is horizontal, but everything can be modified to accommodate a tilt in the launch angle (see the chapter cover figure and Exploration 5.4, Problem 1).

References   Halliday, D., and Resnick, R., *Physics*, (1994: John Wiley & Sons, Inc.)

True, Ernest, "The flight of a ski jumper" in *C·ODE·E*, Spring 1993, pp. 5–8, http://www.math.hmc.edu/codee

Answer questions in the space provided, or on
attached sheets with carefully labeled graphs. A
notepad report using the Architect is OK, too.

Name/Date _____

Course/Section _____

Exploration 5.1. Dunk Tank

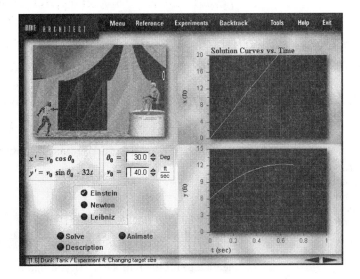

1. *How big is the target?*
    Play the dunk tank game on Screen 1.6 of Module 5 and use various launch
    angles and speeds to help you determine the heights and diameters of the Ein-
    stein, Leibniz, and Newton targets, given that the ball has a 4-inch diameter.

2. *One speed, two angle ranges for success.*
    Use the ODE Architect tool to find two quite different launch angles that will
    dunk Einstein if the launch speed is 40 ft/sec. Repeat for Leibniz and Newton.

3. *Launch angles and speeds that dunk Einstein.*

   Find the region in the $\theta_0 v_0$ plane for which the ball hits the target and dunks Einstein. *Hint*: Start with $v_0 = 40$ ft/sec and determine the ranges for θ_0 using ODE Architect by playing the dunk tank game. Then repeat for other values of v_0.

4. *Solution formulas for the dunk tank model.*

   The position and velocity of the ball at time t is given by formula (2). Find a formula that relates the launch angle to the initial speed and the time T needed to hit the bull's eye. If you had to choose between using your formula and using ODE Architect computer simulations to find winning combinations of speed and launch angle, which would you choose? Why?

Answer questions in the space provided, or on attached sheets with carefully labeled graphs. A notepad report using the Architect is OK, too.

Name/Date _____

Course/Section _____

Exploration 5.2. Longer to Rise or to Fall?

1. *Throw a ball up in the air.*
   Do just that, and determine as best you can the time it takes to rise and to fall. You can use a whiffle ball for slower motion. Explain your results. (No computers here, and no math, either!)

2. *Longer to rise or to fall in a vacuum?*
   What if there were no air to slow the ball down? Use ODE Architect to determine whether it takes the ball longer to rise or to fall. Try various initial speeds between 5 and 60 ft/sec. [*Suggestion:* Use the Sweep feature.]

3. *Longer to rise or to fall with viscous drag?*

    Suppose that air exerts a viscous drag force on a whiffle ball (a reasonable assumption). For various initial speeds, use the ODE Architect to determine whether it takes longer to rise or to fall. Does your answer depend on the initial speed? What physical explanation can you give for your results?

4. *Longer to rise or to fall with your own drag?*

    Repeat Problem 3, but make up several of your own formulas for the drag force. Include Newtonian drag as one case. This isn't as outlandish an idea as it may seem, since the drag force depends very much on the nature of the moving body, e.g., rough or smooth surface, holes through the body, and so on. Discuss your results.

Answer questions in the space provided, or on
attached sheets with carefully labeled graphs. A
notepad report using the Architect is OK, too.

Name/Date _____

Course/Section _____

Exploration 5.3. Indiana Newton

1. *Indiana Newton lands on the boxcar (no drag).*
    Indiana Newton jumps from a height h of 100 ft and intends to land on the
    boxcar of a train moving at a speed of 30 ft/sec. Assuming that there is no air
    resistance, use Screen 2.6 of Module 5 to find the time window of opportunity
    for jumping from the ledge.

2. *Indiana Newton lands on the boxcar (Newtonian drag).*
    Repeat Problem 1 but with Newtonian drag (coefficient $k/m = 0.05$ ft^{-1}).
    Compare fall-times with the no-drag and also with the viscous-drag ($k/m =
    0.05$ sec^{-1}) cases. Find nonzero values of the coefficients so that Indiana
    Newton hits the train sooner with Newtonian drag than with viscous drag.
    How do the fall-times change as Indy's jump height h varies?

3. *How long is the boxcar?*

Use computer simulations of Indiana Newton jumping onto the boxcar and estimate the length of the boxcar.

4. *Indiana Newton floats down.*

First, explore Indiana's fall-times in the viscous- and Newtonian-drag cases where the coefficient k/m has magnitudes ranging from 0 to 0.5 sec^{-1}. Then find a formula in terms of h, k/m, and t for Indiana's position after he jumps: first in the viscous-drag case, then in the Newtonian-drag case. *Suggestion*: In the viscous case, first solve $v' = -32 - kv/m$, $v(0) = 0$, and then integrate and use $y(0) = h$ to get $y(t)$. In the Newtonian-drag case proceed similarly but with $v' = -32 - kv^2$. (This one is hard!) Choose values of the parameters in each case, and compare the graphs of the height function $y(t)$ from your formula with the graphs obtained by the ODE Architect. Any differences?

Answer questions in the space provided, or on attached sheets with carefully labeled graphs. A notepad report using the Architect is OK, too.

Name/Date ⸻

Course/Section ⸻

Exploration 5.4. Ski Jumping

1. *Tilt the edge of the ski jump upward.*
    Use the first Things-to-Think-About on Screen 3.7 of Module 5 to see what happens to the ski jumper's path if the edge of the ski jump structure is tilted upward at 0.524 radians (about 30°). Set $a = 0.01$ sec^{-1}, and $v_0 = 85$ ft/sec, and sweep on the lift coefficient b from 0 to 1.0 in 20 steps. Compare your graphs of the jumper's path with the chapter cover figure. Then animate your graphs. Now fix b at the value 0.02 sec^{-1} and sweep on θ (in radians) to see the effect of the tilt angle on the jumper's path. Explain your results.

2. *Loop-the-loop.*
    The second Things-to-Think-About on Screen 3.7 of Module 5 asks you to use the ODE Architect to estimate the smallest value of the viscous damping coefficient b that will allow the ski jumper to loop-the-loop. If $a = 0.01$ sec^{-1} and $v_0 = 85$ ft/sec, estimate that value. Then increase b by increments from that value upward all the way to the unrealistic value of 5.0 sec^{-1} and describe what you see.

3. *Complex eigenvalues and loop-the-loops.*

☞ You need to know about matrices and eigenvalues to complete this part. See also Chapter 6.

The system matrix of the viscous drag/lift model for ski jumping is

$$\begin{bmatrix} -a & -b \\ b & -a \end{bmatrix}$$

Explain why the eigenvalues of this matrix are complex conjugates with negative real parts if a and b are any positive real numbers. Explain why you are more likely to see loop-the-loops if a is small and b is large. Do some simulations with the ODE Architect for various values of a and b that support your explanation. If you plot a loop-the-loop path over a long enough time interval, you will see no loops at all near the end of the interval. Any explanation?

4. *Newton on skis.*

The fourth Things-to-Think-About on Screen 3.7 of Module 5 puts Indiana Newton on skis with Newtonian drag (of course!). This situation takes you to the expert solver in the ODE Architect, where you are asked to explore every scenario you can think of and to explain what you see in the graphs.

6 First-Order Linear Systems

Oscillating displacements $x_1(t)$ and $x_2(t)$ of two coupled springs play off against each other.

Overview This chapter outlines some of the main facts concerning systems of first-order linear ODEs, especially those with constant coefficients. You'll have the opportunity to work with physical problems that have two or more dependent variables. Such problems can be modeled using systems of differential equations, which can always be written as systems of first-order equations, as can higher-order differential equations. The eigenvalues and eigenvectors of a matrix of coefficients help us understand the behavior of solutions of these systems.

Key words Linear systems; pizza and video; coupled springs; connected tanks; linearized double pendulum; matrix; component; component plot; phase space; phase plane; phase portrait; eigenvalue; eigenvector; saddle point; node; spiral; center; source; sink

See also Chapter 5 for definitions of vector mathematics.

◆ Background

Many applications involve a single independent variable (usually time) and two or more dependent variables. Some examples of dependent variables are:

- the concentrations of a chemical in organs of the body
- the voltage drops across the elements of an electrical network
- the populations of several interacting species
- the profits of businesses in a mall

Applications with more than one dependent variable lead naturally to *systems* of ordinary differential equations. Such systems, as well as higher-order ODEs, can be rewritten as systems of first-order ODEs.

☞ How to convert a second-order ODE to a system of first-order ODEs.

Here's how to reduce a second-order ODE to a system of first-order ODEs (see also Chapter 4). Let's look at the the second-order ODE

$$y'' = f(t, y, y') \qquad (1)$$

Introduce the variables $x_1 = y$ and $x_2 = y'$. Then we get the first-order system

$$x_1' = x_2 \qquad (2)$$
$$x_2' = f(t, x_1, x_2) \qquad (3)$$

ODE (2) follows from the definition of x_1 and x_2, and ODE (3) is ODE (1) rewritten in terms of x_1 and x_2.

✓ "Check" your understanding now by reducing the second-order ODE $y'' + 5y' + 4y = 0$ to a system of first-order ODEs.

◆ Examples of Systems: Pizza and Video, Coupled Springs

Module 6 shows how to model the profits $x(t)$ and $y(t)$ of a pizza parlor and a video store by a system that looks like this:

$$x' = ax + by + c$$
$$y' = fx + gy + h$$

where a, b, c, f, g, and h are constants. Take another look at Screens 1.1–1.4 in Module 6 to see how ODE Architect handles these systems.

Module 6 also presents a model system of second-order ODEs for oscillating springs and masses. A pair of coupled springs with spring constants k_1 and k_2 are connected to masses m_1 and m_2 that glide back forth on a table. As shown in the "Coupled Springs" submodule, if damping is negligible then the second-order linear ODEs that model the displacements of the masses from equilibria are

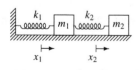

$$m_1 x_1'' = -(k_1 + k_2)x_1 + k_2 x_2$$
$$m_2 x_2'' = k_2 x_1 - k_2 x_2$$

☞ A system of first-order ODEs is *autonomous* if the terms on the right-hand sides of the equations do not explicitly depend on time.

Let's set $m_1 = 4$, $m_2 = 1$, $k_1 = 3$, and $k_2 = 1$. Then, setting $x_1' = v_1$, $x_2' = v_2$, the corresponding autonomous system of four first-order ODEs is

$$x_1' = v_1$$
$$v_1' = -x_1 + \frac{1}{4}x_2$$
$$x_2' = v_2$$
$$v_2' = x_1 - x_2$$

☞ Trajectories of an autonomous system can't intersect because to do so would violate the uniqueness property that only one trajectory can pass through a given point.

The cover figure of this chapter shows how x_1 and x_2 play off against each other when $x_1(0) = 0.4$, $v_1(0) = 1$, $x_2(0) = 0$, and $v_2(0) = 0$. The trajectories for this IVP are defined in the 4-dimensional $x_1 v_1 x_2 v_2$-space and cannot intersect themselves. However, the projections of the trajectories onto any plane *can* intersect, as we see in the cover figure.

◆ Linear Systems with Constant Coefficients

The model first-order systems of ODEs for pizza and video and for coupled springs have the special form of linear systems with constant coefficients. Now we shall see just what linearity means and how it allows us (sometimes) to construct solution formulas for linear systems.

Let t (time) be the independent variable and let $x_1, x_2, \ldots, x_n$ denote the dependent variables. Then a general system of first-order linear *homogeneous* ODEs with constant coefficients has the form

☞ Dependent variables are also called *state variables*.

☞ *Homogeneous* means that there are no free terms, that is, terms that don't involve any x_i.

$$\begin{aligned} x_1' &= a_{11}x_1 + a_{12}x_2 + \cdots + a_{1n}x_n \\ x_2' &= a_{21}x_1 + a_{22}x_2 + \cdots + a_{2n}x_n \\ &\vdots \\ x_n' &= a_{n1}x_1 + a_{n2}x_2 + \cdots + a_{nn}x_n \end{aligned} \tag{4}$$

where $a_{11}, a_{12}, \ldots, a_{nn}$ are given constants. To find a unique solution, we need a set of initial conditions, one for each dependent variable:

$$x_1(t_0) = \alpha_1, \quad \ldots, \quad x_n(t_0) = \alpha_n \tag{5}$$

where t_0 is a specific time and $\alpha_1, \ldots, \alpha_n$ are given constants. The system (4) and the initial conditions (5) together constitute an *initial value problem* (IVP) for $x_1, \ldots, x_n$ as functions of t. Note that $x_1 = \cdots = x_n = 0$ is an equilibrium point of system (4).

☞ An equilibrium point of an autonomous system of ODEs is a point where all the rates are zero; it corresponds to a constant solution.

The model on Screen 1.4 of Module 6 for the profits of the pizza and video stores is the system

$$\begin{aligned} x' &= 0.06x + 0.01y - 0.013 \\ y' &= 0.04x + 0.05y - 0.013 \end{aligned} \tag{6}$$

☞ If $n = 2$, we often use x and y for the dependent variables.

with the initial conditions

$$x(0) = 0.30, \quad y(0) = 0.20 \tag{7}$$

The ODEs (6) are nonhomogeneous due to the presence of the free term -0.013 in each equation. The coordinates of an *equilibrium point* of a system are values of the dependent variables for which all of the derivatives $x_1', \ldots, x_n'$ are zero. For the system (6) the only equilibrium point is $(0.2, 0.1)$. The translation $X = x - 0.2$, $Y = y - 0.1$ transforms the system (6) into the system

☞ A change of variables puts the equilibrium point at the origin.

$$X' = 0.06X + 0.01Y$$
$$Y' = 0.04X + 0.05Y \tag{8}$$

which is homogeneous and has the same coefficients as the system (6). In terms of X and Y, the initial conditions (7) become

$$X(0) = 0.1, \quad Y(0) = 0.1 \tag{9}$$

Although we have converted a nonhomogeneous system to a homogeneous system in this particular case, it isn't always possible to do so.

☞ Vectors and matrices appear as bold letters.

☞ **A** is called the *linear system matrix*, or the *Jacobian matrix*.

It is useful here to introduce matrix notation: it saves space and it expresses system (4) in the form of a single equation. Let **x** be the vector with components $x_1, x_2, \ldots, x_n$ and let **A** be the matrix of the coefficients, where a_{ij} is the element in the ith row and jth column of **A**. The derivative of the vector **x**, written $d\mathbf{x}/dt$, or $\mathbf{x}'$ is defined to be the vector with the components $dx_1/dt, \ldots, dx_n/dt$. Therefore we can write the system (4) in the compact form

☞ The vector $\mathbf{x}(t)$ is called the *state* of system (10) at time t.
$$\mathbf{x} = \begin{bmatrix} x_1 \\ \vdots \\ x_n \end{bmatrix}, \quad \mathbf{x}' = \begin{bmatrix} x_1' \\ \vdots \\ x_n' \end{bmatrix}$$

$$\mathbf{x}' = \mathbf{A}\mathbf{x}, \quad \text{where} \quad \mathbf{A} = \begin{bmatrix} a_{11} & \cdots & a_{1n} \\ \vdots & & \vdots \\ a_{n1} & \cdots & a_{nn} \end{bmatrix} \tag{10}$$

In vector notation, the initial conditions (5) become

$$\mathbf{x}(t_0) = \alpha \tag{11}$$

where α is the vector with components $\alpha_1, \ldots, \alpha_n$.

✓ Find the linear system matrix for system (8).

A solution of the initial value problem (10) and (11) is a set of functions

$$x_1 = x_1(t)$$
$$\vdots \tag{12}$$
$$x_n = x_n(t)$$

that satisfy the differential equations and initial conditions. Using our new notation, if $\mathbf{x}(t)$ is the vector whose components are $x_1(t), \ldots, x_n(t)$, then $\mathbf{x} = \mathbf{x}(t)$ is a solution of the corresponding vector IVP, (10) and (11). The system $\mathbf{x}' = \mathbf{A}\mathbf{x}$ is homogeneous, while a nonhomogeneous system would have the form $\mathbf{x}' = \mathbf{A}\mathbf{x} + \mathbf{F}$, where **F** is a vector function of t or else a constant vector.

◆ Solution Formulas: Eigenvalues and Eigenvectors

To find a solution formula for system (10) let's look for an exponential solution of the form

$$\mathbf{x} = \mathbf{v}e^{\lambda t} \tag{13}$$

where λ is a constant and $\mathbf{v}$ is a constant vector to be determined. Substituting $\mathbf{x}$ as given by (13) into the ODE (10), we find that $\mathbf{v}$ and λ must satisfy the algebraic equation

$$\mathbf{A}\mathbf{v} = \lambda\mathbf{v} \tag{14}$$

Equation (14) can also be written in the form

$$(\mathbf{A} - \lambda\mathbf{I})\mathbf{v} = \mathbf{0} \tag{15}$$

where $\mathbf{I}$ is the *identity matrix* and $\mathbf{0}$ is the *zero vector* with zero for each component. Equation (15) has nonzero solutions if and only if λ is a root of the nth-degree polynomial equation

☞ The determinant of a matrix is denoted by det.

$$\det(\mathbf{A} - \lambda\mathbf{I}) = 0 \tag{16}$$

called the *characteristic equation* for the system (10). Such a root is called an *eigenvalue* of the matrix $\mathbf{A}$. We will denote the eigenvalues by $\lambda_1, \ldots, \lambda_n$. For each eigenvalue λ_i there is a corresponding nonzero solution $\mathbf{v}^{(i)}$, called an *eigenvector*. The eigenvectors are not determined uniquely but only up to an arbitrary multiplicative constant.

☞ The keys to finding a solution formula for $\mathbf{x}' = \mathbf{A}\mathbf{x}$ are the eigenvalues and eigenvectors of $\mathbf{A}$.

For each eigenvalue-eigenvector pair $(\lambda_i, \mathbf{v}^{(i)})$ there is a corresponding vector solution $\mathbf{v}^{(i)}e^{\lambda_i t}$ of the ODE (10). If the eigenvalues $\lambda_1, \ldots, \lambda_n$ are all different, then there are n such solutions,

$$\mathbf{v}^{(1)}e^{\lambda_1 t}, \ldots, \mathbf{v}^{(n)}e^{\lambda_n t}$$

☞ Formula (17) is called the general solution formula of system (10) because every solution has the form of (17) for some choice of the constants C_j. The other way around, every choice of the constants yields a solution of system (10).

In this case the *general solution* of system (10) is the linear combination

$$\mathbf{x} = C_1\mathbf{v}^{(1)}e^{\lambda_1 t} + \cdots + C_n\mathbf{v}^{(n)}e^{\lambda_n t} \tag{17}$$

The arbitrary constants $C_1, \ldots, C_n$ can always be chosen to satisfy the initial conditions (11). If the eigenvalues are not distinct, then the general solution takes on a slightly different (but similar) form. The texts listed in the references give the formulas for this case. If some of the eigenvalues are complex, then the solution given by formula (17) is complex-valued. However, if all of the coefficients a_{ij} are real, then the complex eigenvalues and eigenvectors occur in complex conjugate pairs, and it is always possible to express the solution formula (17) in terms of real-valued functions. Look ahead to formulas (20) and (21) for a way to accomplish this feat.

◆ Calculating Eigenvalues and Eigenvectors

Here's how to find the eigenvalues and eigenvectors of a 2×2 real matrix

$$\mathbf{A} = \begin{bmatrix} a & b \\ c & d \end{bmatrix}$$

First define the *trace* of $\mathbf{A}$ (denoted by $\operatorname{tr}\mathbf{A}$) to be the sum $a + d$ of the diagonal entries, and the *determinant* of $\mathbf{A}$ (denoted by $\det\mathbf{A}$) to be the number $ad - bc$. Then the characteristic equation for $\mathbf{A}$ is

$$\begin{aligned}
\det(\mathbf{A} - \lambda\mathbf{I}) &= \det\begin{bmatrix} a - \lambda & b \\ c & d - \lambda \end{bmatrix} \\
&= \lambda^2 - (a + d)\lambda + ad - bc \\
&= \lambda^2 - (\operatorname{tr}\mathbf{A})\lambda + \det\mathbf{A} \\
&= 0
\end{aligned}$$

The eigenvalues of $\mathbf{A}$ are the roots λ_1 and λ_2 of this quadratic equation. We assume $\lambda_1 \neq \lambda_2$. For the eigenvalue λ_1 we can find a corresponding eigenvector $\mathbf{v}^{(1)}$ by solving the vector equation

$$\mathbf{A}\mathbf{v}^{(1)} = \lambda_1\mathbf{v}^{(1)}$$

for $\mathbf{v}^{(1)}$. In a similar fashion we can find an eigenvector $\mathbf{v}^{(2)}$ corresponding to the eigenvalue λ_2.

Example: Take a look at the system

$$\mathbf{x}' = \mathbf{A}\mathbf{x}, \quad \mathbf{A} = \begin{bmatrix} 0 & 1 \\ -2 & 3 \end{bmatrix}, \quad \mathbf{x} = \begin{bmatrix} x_1 \\ x_2 \end{bmatrix} \tag{18}$$

Since

$$\operatorname{tr}\mathbf{A} = 0 + 3 = 3 \quad \text{and} \quad \det\mathbf{A} = 0 \cdot 3 - 1 \cdot (-2) = 2$$

the characteristic equation is

$$\lambda^2 - (\operatorname{tr}\mathbf{A})\lambda + \det\mathbf{A} = \lambda^2 - 3\lambda + 2 = 0$$

The eigenvalues are $\lambda_1 = 1$ and $\lambda_2 = 2$. To find an eigenvector $\mathbf{v}^{(1)}$ for λ_1, let's solve

$$\begin{bmatrix} 0 & 1 \\ -2 & 3 \end{bmatrix}\mathbf{v}^{(1)} = \lambda_1\mathbf{v}^{(1)} = \mathbf{v}^{(1)}$$

for $\mathbf{v}^{(1)}$. Denoting the components of $\mathbf{v}^{(1)}$ by α and β, we have

$$\begin{bmatrix} 0 & 1 \\ -2 & 3 \end{bmatrix}\begin{bmatrix} \alpha \\ \beta \end{bmatrix} = \begin{bmatrix} \beta \\ -2\alpha + 3\beta \end{bmatrix} = \begin{bmatrix} \alpha \\ \beta \end{bmatrix}$$

This gives two equations for α and β:

$$\beta = \alpha, \quad -2\alpha + 3\beta = \beta$$

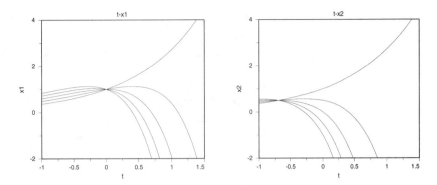

Figure 6.1: Graphs of five solutions $x_1(t)$ (left), $x_2(t)$ (right) of system (18).

The second equation is equivalent to the first, so we may as well set $\alpha = \beta = 1$, which gives us an eigenvector $\mathbf{v}^{(1)}$. In a similar way for the eigenvalue λ_2, we can find an eigenvector $\mathbf{v}^{(2)}$ with components $\alpha = 1$, $\beta = 2$. So the general solution of $\mathbf{x}' = \mathbf{A}\mathbf{x}$ in this case is

$$\mathbf{x} = C_1 \mathbf{v}^{(1)} e^{\lambda_1 t} + C_2 \mathbf{v}^{(2)} e^{\lambda_2 t}$$

$$= C_1 \begin{bmatrix} 1 \\ 1 \end{bmatrix} e^t + C_2 \begin{bmatrix} 1 \\ 2 \end{bmatrix} e^{2t}$$

or in component form

$$x_1 = C_1 e^t + C_2 e^{2t}$$
$$x_2 = C_1 e^t + 2C_2 e^{2t}$$

where C_1 and C_2 are arbitrary constants.

✓ Find a formula for the solution of system (18) if $x_1(0) = 1$, $x_2(0) = -1$. Figure 6.1 shows graphs of $x_1(t)$ and $x_2(t)$ where $x_1(0) = 1$, $x_2(0) = 0, \pm 0.5$, ± 1. Which graphs correspond to $x_1(0) = 1$, $x_2(0) = -1$? What happens as $t \to +\infty$? As $t \to -\infty$?

◆ Phase Portraits

We can view solutions graphically in several ways. For example, we can draw plots of $x_1(t)$ vs. t, $x_2(t)$ vs. t, and so on. These plots are called *component plots* (see Figure 6.1). Alternatively, we can interpret equations (12) as a set of parametric equations with t as the parameter. Then each specific value of t corresponds to a set of values for $x_1, \ldots, x_n$. We can view this set of values as coordinates of a point in $x_1 x_2 \cdots x_n$-space, called the *phase space*. (If $n = 2$ it's called the *phase plane*.) For an interval of t-values, the corresponding points form a curve in phase space. This curve is called a *phase plot*, a *trajectory*, or an *orbit*.

☞ Another term for phase space is *state space*.

Phase plots are particularly useful if $n = 2$. In this case it is often worthwhile to draw several trajectories starting at different initial points on the same set of axes. This produces a *phase portrait*, which gives us the best possible overall view of the behavior of solutions. Whatever the value of n, the trajectories of system (10) can never intersect because system (10) is autonomous.

If **A** in system (10) is a 2×2 matrix, then it is useful to examine and classify the various cases that can arise. There aren't many cases when $n = 2$, but even so these cases give important information about higher-dimensional linear systems, as well as nonlinear systems (see Chapter 7). We won't consider here the cases where the two eigenvalues are equal, or where one or both of them are zero.

A *direction field* (or *vector field*) for an autonomous system when $n = 2$ is a field of line segments. The slope of the segment at the point (x_1, x_2) is x_2'/x_1'. The trajectory through (x_1, x_2) is tangent to the segment. An arrowhead on the segment shows the direction of the flow. See Figures 6.2–6.5 for examples.

Real Eigenvalues
If the eigenvalues λ_1 and λ_2 are real, the general solution is

$$\mathbf{x} = C_1 \mathbf{v}^{(1)} e^{\lambda_1 t} + C_2 \mathbf{v}^{(2)} e^{\lambda_2 t} \tag{19}$$

where C_1 and C_2 are arbitrary real constants.

☞ Trajectories starting on either line at $t = 0$ stay on the line.

Let's first look at the case where λ_1 and λ_2 have opposite signs, with $\lambda_1 > 0$ and $\lambda_2 < 0$. The term in formula (19) involving λ_1 dominates as $t \to +\infty$, and the term involving λ_2 dominates as $t \to -\infty$. Thus as $t \to +\infty$ the trajectories approach the line that goes through the origin and has the same slope as $\mathbf{v}^{(1)}$, and as $t \to -\infty$, they approach the line that goes through the origin and has the same slope as $\mathbf{v}^{(2)}$. A typical phase portrait for this case is shown in Figure 6.2. The origin is called a *saddle point*, and it is *unstable*, since most solutions move away from the point.

☞ Eigenvalues of opposite signs imply a *saddle*.

Now suppose that λ_1 and λ_2 are both negative, with $\lambda_2 < \lambda_1 < 0$. The solution is again given by formula (19), but in this case both terms approach

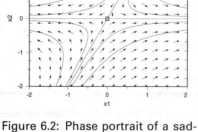

Figure 6.2: Phase portrait of a saddle: $x_1' = x_1 - x_2,\ x_2' = -x_2$.

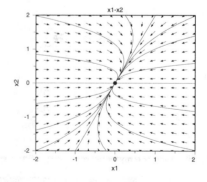

Figure 6.3: Phase portrait of a nodal sink: $x_1 = -3x_1 + x_2,\ x_2' = -x_2$.

zero as $t \to +\infty$. However, for large positive t, the factor $e^{\lambda_2 t}$ is much smaller than $e^{\lambda_1 t}$, so for $C_1 \neq 0$ the trajectories approach the origin tangent to the line with the same slope as $\mathbf{v}^{(1)}$, and if $C_1 = 0$ the trajectory lies on the line with the same slope as $\mathbf{v}^{(2)}$. For large negative t, the term involving λ_2 is the dominant one and the trajectories approach asymptotes that have the same slope as $\mathbf{v}^{(2)}$. A typical phase portrait for this case is shown in Figure 6.3. The origin attracts all solutions and is called an *asymptotically stable node*. It is also called a *sink* because all nearby orbits get pulled in as $t \to +\infty$.

☞ Both eigenvalues negative imply a *nodal sink*.

If both eigenvalues are positive, the situation is similar to when both eigenvalues are negative, but in this case the direction of motion on the trajectories is reversed. For example, suppose that $0 < \lambda_1 < \lambda_2$: then the trajectories are unbounded as $t \to +\infty$ and asymptotic to lines parallel to $\mathbf{v}^{(2)}$. As $t \to -\infty$ the trajectories approach the origin either tangent to the line through the origin with the same slope as $\mathbf{v}^{(1)}$ or lying on the line through the origin with the same slope as $\mathbf{v}^{(2)}$. A typical phase portrait for this case looks like Figure 6.3 but with the arrows reversed. The origin is an *unstable node*. It is also called a *source* because all orbits (except $\mathbf{x} = \mathbf{0}$ itself) flow out and away from the origin as t increases from $-\infty$.

☞ Both eigenvalues positive imply a *nodal source*.

✓ Find the eigenvalues and eigenvectors of the systems of Figures 6.2 and 6.3 and interpret them in terms of the phase plane portraits.

Complex Eigenvalues

Now suppose that the eigenvalues are complex conjugates $\lambda_1 = \alpha + i\beta$ and $\lambda_2 = \alpha - i\beta$. The exponential form (13) of a solution remains valid, but usually it is preferable to use Euler's formula:

$$e^{i\beta t} = \cos(\beta t) + i \sin(\beta t) \tag{20}$$

This allows us to write the solution in terms of real-valued functions. The result is

$$\mathbf{x} = C_1 e^{\alpha t}[\mathbf{a}\cos(\beta t) - \mathbf{b}\sin(\beta t)] + C_2 e^{\alpha t}[\mathbf{b}\cos(\beta t) + \mathbf{a}\sin(\beta t)] \tag{21}$$

where $\mathbf{a}$ and $\mathbf{b}$ are the real and imaginary parts of the eigenvector $\mathbf{v}^{(1)}$ associated with λ_1, and C_1 and C_2 are constants. The trajectories are spirals about the origin. If $\alpha > 0$, then the spirals grow in magnitude and the origin is called a *spiral source* or an *unstable spiral point*. A typical phase portrait in this case looks like Figure 6.4. If $\alpha < 0$, then the spirals approach the origin as $t \to +\infty$, and the origin is called a *spiral sink* or an *asymptotically stable spiral point*. In both cases the spirals encircle the origin and may be directed in either the clockwise or counterclockwise direction (but not both directions in the same system).

☞ Complex eigenvalues with nonzero real parts imply a *spiral sink* or a *spiral source*.

Finally, consider the case $\lambda = \pm i\beta$, where β is real and positive. Now the exponential factors in solution formula (21) are absent so the trajectory is bounded as $t \to \pm\infty$, but it does not approach the origin. In fact, the

☞ Pure imaginary
eigenvalues imply a *center*.

trajectories are ellipses centered on the origin (see Figure 6.5), and the origin is called a *center*. It is *stable*, but not asymptotically stable.

✓   Find the eigenvalues of the systems of Figures 6.4 and 6.5, and interpret them in terms of the phase plane portraits. Why can't you "see" the eigenvectors in these portraits?

There is one other graphing technique that is often useful. If $n = 2$, ODE Architect can draw a plot of the solution in tx_1x_2-space. If we project this curve onto each of the coordinate planes, we obtain the two component plots and the phase plot (Figure 6.6).

◆ Using ODE Architect to Find Eigenvalues and Eigenvectors

ODE Architect will find equilibrium points of a system and the eigenvalues and eigenvectors of the Jacobian matrix of an autonomous system at an equilibrium point. Here are the steps:

- Enter an autonomous system of first-order ODEs.
- Click on the lower left Equilibrium tab; enter a guess for the coordinates of an equilibrium point.
- The Equil. tab at the lower right will bring up a window with calculated coordinates of an equilibrium point close to your guess.

☞ Use this Architect feature
to calculate the eigenvalues,
eigenvectors.

- Double click anywhere on the boxed coordinates of an equilibrium in the window (or click on the window's editing icon) to see the eigenvalues, eigenvectors, and the Jacobian matrix.

If you complete these steps for a system of two first-order, autonomous ODEs, ODE Architect will insert a symbol at the equilibrium point in the phase plane: An open square for a saddle, a solid dot for a sink, an open dot for a source, and a plus sign for a center (Figures 6.2–6.5). The symbols can be edited using the Equilibrium tab on the edit window.

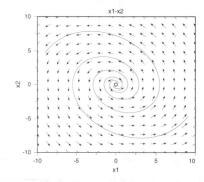

Figure 6.4: Phase portrait of a spiral source: $x_1' = x_2$, $x_2' = x_1 + 0.4x_2$.

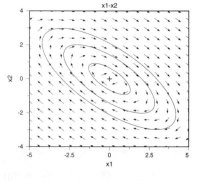

Figure 6.5: Phase portrait of a center: $x_1' = x_1 + 2x_2$, $x_2' = -x_1 - x_2$.

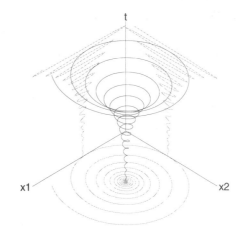

Figure 6.6: Solution curve of $x_1' = x_2$, $x_2' = -100.25x_1 + x_2$, $x_1(0) = 1$, $x_2(0) = 1$, the two component curves, and the trajectory in the x_1x_2-phase plane.

✓ Use ODE Architect to find the eigenvalues and eigenvectors of the system in Figure 6.2.

◆ Separatrices

A trajectory Γ of a planar autonomous system is a *separatrix* if the long-term behavior of trajectories on one side of Γ is quite different from the behavior of those on the other side. Take a look at the four *saddle separatrices* in Figure 6.2, each of which is parallel to an eigenvector of the system matrix. The two separatrices that approach the saddle point as t increases are the *stable separatrices*, and the two that leave are the *unstable separatrices*.

◆ Parameter Movies

The eigenvalues of a 2×2 matrix $\mathbf{A}$ depend on the values of $\operatorname{tr}\mathbf{A}$ and $\det\mathbf{A}$, and the behavior of the trajectories of $\mathbf{x}' = \mathbf{A}\mathbf{x}$ depends very much on the eigenvalues. So it makes sense to see what happens to trajectories as we vary the values of $\operatorname{tr}\mathbf{A}$ and $\det\mathbf{A}$. When we do this varying, we can make the eigenvalues change sign, or move into the complex plane, or become equal. As the changes occur the behavior of the trajectories has to change as well. Take a look at the "Parameter Movies" part of Module 6 for some surprising views of the changing phase plane portraits as we follow along a path in the parameter plane of $\operatorname{tr}\mathbf{A}$ and $\det\mathbf{A}$.

References   Borrelli, R.L., and Coleman, C.S., *Differential Equations: A Modeling Perspective* (1998: John Wiley & Sons, Inc.)

Boyce, W.E., and DiPrima, R.C., *Elementary Differential Equations and Boundary Value Problems*, 6th ed., (1997: John Wiley & Sons, Inc.)

Answer questions in the space provided, or on attached sheets with carefully labeled graphs. A notepad report using the Architect is OK, too.

Name/Date _____

Course/Section _____

Exploration 6.1. Eigenvalues, Eigenvectors, and Graphs

1. Each of the phase portraits in the graphs below is associated with a planar autonomous linear system with equilibrium point at the origin. What can you say about the eigenvalues of the system matrix $\mathbf{A}$ (e.g., are they real, complex, positive)? Sketch by hand any straight line trajectories. What can you say about the eigenvectors of $\mathbf{A}$?

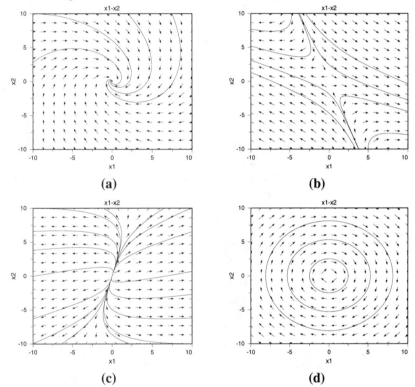

2. What does the phase portrait of $\mathbf{x}' = \mathbf{Ax}$ look like if $\mathbf{A}$ is a 2×2 matrix with one eigenvalue zero and the other nonzero? How many equilibrium points are there? Include portraits of specific examples.

3. Using Figure 6.6 as a guide, make your own gallery of 2D and 3D graphs to illustrate solution curves, component curves, trajectories, and phase-plane portraits of the systems $\mathbf{x}' = \mathbf{Ax}$, where $\mathbf{A}$ is a 2×2 matrix of constants. List eigenvalues and eigenvectors of $\mathbf{A}$. Include examples of the following types of equilibrium points:

   • Saddle
   • Nodal sink
   • Nodal source
   • Spiral sink
   • Spiral source
   • Center
   • Eigenvalues of $\mathbf{A}$ are equal and negative

Answer questions in the space provided, or on attached sheets with carefully labeled graphs. A notepad report using the Architect is OK, too.

Name/Date _____

Course/Section _____

Exploration 6.2. Pizza and Video

Sometimes business enterprises are strongly affected by periodic (e.g., seasonal) influences. We can illustrate this in the case of Diffey and Cue.

The model describing Diffey's and Cue's profits on Screen 1.4 in Module 6 is

$$x' = 0.06x + 0.01y - 0.013$$
$$y' = 0.04x + 0.05y - 0.013$$

(22)

Let's introduce a periodic fluctuation in the coefficient of x in the first ODE and in the coefficient of y in the second ODE.

Sine and cosine functions are often used to model periodic phenomena. We'll use $\sin(2\pi t)$ so that the fluctuations have a period of one time unit. We will also include a variable amplitude parameter a so that the intensity of the fluctuations can be easily controlled. We have the modified system

$$x' = 0.06 \left(1 + \frac{1}{2}a\sin(2\pi t)\right)x + 0.01y - 0.013$$

$$y' = 0.04x + 0.05 \left(1 + \frac{3}{10}a\sin(2\pi t)\right)y - 0.013$$

(23)

Note that if $a = 0$, we recover system (22), and that as a increases the amplitude of the fluctuations in the coefficients also increases.

1. Interpret the terms involving $\sin(2\pi t)$ in the context of Diffey's and Cue's businesses. Use ODE Architect to solve the system (23) subject to the initial conditions $x(0) = 0.3$, $y(0) = 0.2$ for $a = 1$. Use the time interval $0 \le t \le 10$, or an even longer interval. Plot x vs. t, y vs. t, and y vs. x. Compare the plots with the corresponding plots for the system (22). What is the effect of the fluctuating coefficients on the solution? Repeat with the same initial data, but sweeping a from 0 to 5 in 11 steps. What is the effect of increasing a on the solution?

2. Use ODE Architect to solve the system (23) subject to the initial conditions $x(0) = 0.25$, $y(0) = 0$ for $a = 3$. Draw a plot of y vs. x only. Be sure to use a sufficiently large t-interval to make clear the ultimate behavior of the solution. Repeat using the initial conditions $x(0) = 0.2$, $y(0) = -0.2$. Explain what you see.

3. For the two initial conditions in Problem 2 you should have found solutions that behave quite differently. Consider initial points on the line joining $(0.25, 0)$ and $(0.2, -0.2)$. For $a = 3$, estimate the coordinates of the point where the solution changes from one type of behavior to the other.

Answer questions in the space provided, or on
attached sheets with carefully labeled graphs. A
notepad report using the Architect is OK, too.

Name/Date _____

Course/Section _____

Exploration 6.3. Control of Interconnected Water Tanks

☞ Take a look at
Chapter 8 for a way to
diagram this
"compartment" model.

Consider two interconnected tanks containing salt water. Initially Tank 1 contains 5 gal of water and 3 oz of salt while Tank 2 contains 4 gal of water and 5 oz of salt.

Water containing p_1 oz of salt per gal flows into Tank 1 at a rate of 2 gal/min. The mixture in Tank 1 flows out at a rate of 6 gal/min, of which half goes into Tank 2 and half leaves the system.

Water containing p_2 oz of salt per gal flows into Tank 2 at a rate of 3 gal/min. The mixture in Tank 2 flows out at a rate of 6 gal/min: 4 gal/min goes to Tank 1, and the rest leaves the system.

1. Draw a diagram showing the tank system. Does the amount of water in each tank remain the same during this flow process? Explain. If $q_1(t)$ and $q_2(t)$ are the amounts of salt (in oz) in the respective tanks at time t, show that they satisfy the system of differential equations:

$$q_1' = 2p_1 - \tfrac{6}{5}q_1 + q_2$$
$$q_2' = 3p_2 + \tfrac{3}{5}q_1 - \tfrac{3}{2}q_2$$

What are the initial conditions associated with this system of ODEs?

2. Suppose that $p_1 = 1$ oz/gal and $p_2 = 1$ oz/gal. Solve the IVP, plot $q_1(t)$ vs. t, and estimate the limiting value q_1^* that $q_1(t)$ approaches after a long time. In a similar way estimate the limiting value q_2^* for $q_2(t)$. Repeat for your own initial conditions, but remember that $q_1(0)$ and $q_2(0)$ must be nonnegative. How are q_1^* and q_2^* affected by changes in the initial conditions? Now use ODE Architect to find q_1^* and q_2^*. [*Hint:* Use the Equilibrium tab.] Is the equilibrium point a source or a sink? A node, saddle, spiral, or center?

3. The operator of this system (you) can control it by adjusting the input parameters p_1 and p_2. Note that q_1^* and q_2^* depend on p_1 and p_2. Find values of p_1 and p_2 so that $q_1^* = q_2^*$. Can you find values of p_1 and p_2 so that $q_1^* = 1.5q_2^*$? So that $q_2^* = 1.5q_1^*$?

4. Let c_1^* and c_2^* be the limiting concentrations of salt in each tank. Express c_1^* and c_2^* in terms of q_1^* and q_2^*, respectively. Find p_1 and p_2, if possible, so as to achieve each of the following results:

(a) $c_1^* = c_2^*$ **(b)** $c_1^* = 1.5c_2^*$ **(c)** $c_2^* = 1.5c_1^*$

Finally, consider all possible (nonnegative) values of p_1 and p_2. Describe the set of limiting concentrations c_1^* and c_2^* that can be obtained by adjusting p_1 and p_2.

Answer questions in the space provided, or on attached sheets with carefully labeled graphs. A notepad report using the Architect is OK, too.

Name/Date _____

Course/Section _____

Exploration 6.4. Three Interconnected Tanks

☞ Take a look at Chapter 8 for a way to diagram this "compartment" model.

Consider three interconnected tanks containing salt water. Initially Tanks 1 and 2 contain 10 gal of water while Tank 3 contains 15 gal. Each tank initially contains 6 oz of salt.

Water containing 2 oz of salt per gal flows into Tank 1 at a rate of 1 gal/min. The mixture in Tank 1 flows into Tank 2 at a rate of r gal/min. Furthermore, the mixture in Tank 1 is discharged into the drain at a rate of 2 gal/min. Water containing 1 oz of salt per gal flows into Tank 2 at a rate of 2 gal/min. The mixture in Tank 2 flows into Tank 3 at a rate of $r+1$ gal/min and also flows back into Tank 1 at a rate of 1 gal/min. The mixture in Tank 3 flows into Tank 1 at a rate of r gal/min, and down the drain at a rate of 1 gal/min.

1. Draw a diagram that depicts the tank system. Does the amount of water in each tank remain constant during the process? Show that the flow process is modeled by the following system of equations, where $q_1(t)$, $q_2(t)$, and $q_3(t)$ are the amounts of salt (in oz) in the respective tanks at time t:

$$q_1' = 2 - \frac{r+2}{10}q_1 + \frac{1}{10}q_2 + \frac{r}{15}q_3$$

$$q_2' = 2 + \frac{r}{10}q_1 - \frac{r+2}{10}q_2$$

$$q_3' = \frac{r+1}{10}q_2 - \frac{r+1}{15}q_3$$

What are the corresponding initial conditions?

2. Let $r = 1$, and use ODE Architect to plot q_1 vs. t, q_2 vs. t, and q_3 vs. t for the IVP in Problem 1. Estimate the limiting value of the amount of salt in each tank after a long time. Now suppose that the flow rate r is increased to 4 gal/min. What effect do you think this will have on the limiting values for q_1, q_2, and q_3? Check your intuition with ODE Architect. What do you think will happen to the limiting values if r is increased further? For each value of r use ODE Architect to find the limiting values for q_1, q_2, and q_3.

3. Although the two sets of graphs in Problem 2 may look similar, they're actually slightly different. Calculate the eigenvalues of the coefficient matrix

☞ Use ODE Architect to find the eigenvalues.

when $r = 1$ and when $r = 4$. There is a certain "critical" value $r = r_0$ between 1 and 4 where complex eigenvalues first occur. Determine r_0 to two decimal places.

4. Complex eigenvalues lead to sinusoidal solutions. Explain why the oscillatory behavior characteristic of the sine and cosine functions is not apparent in your graphs from Problem 2 for $r = 4$. Devise a plan that will enable you to construct plots showing the oscillatory part of the solution for $r = 4$. Then execute your plan to make sure that it is effective.

Answer questions in the space provided, or on
attached sheets with carefully labeled graphs. A
notepad report using the Architect is OK, too.

Name/Date _____

Course/Section _____

Exploration 6.5. Small Motions of a Double Pendulum; Coupled Springs

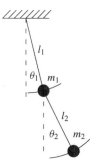

Another physical system with two degrees of freedom is the planar double pendulum. This consists of two rods of length l_1 and l_2 and two masses m_1 and m_2, all attached together so that motions are confined to a vertical plane. Here we'll investigate motions for which the pendulum system doesn't move too far from its stable equilibrium position in which both rods are hanging vertically downward. We'll assume the damping in this system is negligible.

A sketch of the double pendulum system is shown in the margin. A derivation of the nonlinear equations in terms of the angles $\theta_1(t)$ and $\theta_2(t)$ that govern the oscillations of the system is given in Chapter 7 (beginning on page 126). The equations of interest here are the linearized ODEs in θ_1 and θ_2 where both of these angles are required to be small:

$$l_1\theta_1'' + \frac{m_2}{m_1 + m_2}l_2\theta_2'' + g\theta_1 = 0$$

$$l_2\theta_2'' + l_1\theta_1'' + g\theta_2 = 0$$

For small values of θ_1, θ_1', θ_2, and θ_2' these ODEs are obtained by linearizing ODEs (19) and (20) on page 127.

1. Consider the special case where $m_1 = m_2 = m$ and $l_1 = l_2 = l$, and define $g/l = \omega_0^2$. Write the equations above as a system of four first-order equations. Use ODE Architect to generate motions for different values of ω_0. Experiment with different initial conditions. Try to visualize the motions of the pendulum system that correspond to your solutions. Then use the model-based animation tool in ODE Architect and watch the animated double pendulums gyrate as your initial value problems are solved.

2. Assume $\omega_0^2 = 10$ in Problem 1. Can you find in-phase and out-of-phase oscillations that are analogous to those of the coupled mass-spring system? Determine the relationships between the initial conditions $\theta_1(0)$ and $\theta_2(0)$ that are needed to produce these motions. Plot θ_2 against θ_1 for these motions. Then change $\theta_1(0)$ or $\theta_2(0)$ to get a motion which is neither in-phase nor out-of-phase. Overlay this graph on the first plot. Explain what you see. Use the model-based animation feature in ODE Architect to help you "see" the in-phase and out-of-phase motions, and those that are neither. Describe what you see.

3. Show that the linearized equations for the double pendulum in Problem 2 are equivalent to those for a particular coupled mass-spring system. Find the corresponding values of (or constraints on) the mass-spring parameters m_1, m_2, k_1, and k_2. Does this connection extend to other double-pendulum parameter values besides those in Problems 1 and 2? If so, find the relationships between the parameters of the corresponding systems. Use the model-based animation feature in ODE Architect and watch the springs vibrate and the double pendulum gyrate. Describe what you see.

7 Nonlinear Systems

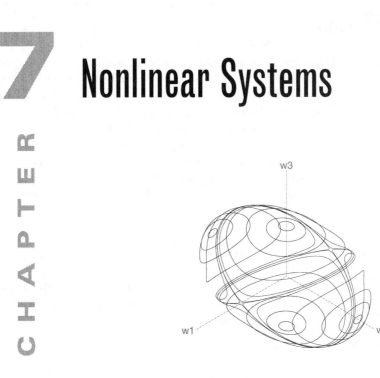

Angular velocity: twenty-four ways to spin a book.

Overview   While many natural processes can be modeled by linear systems of ODEs, others require *nonlinear systems*. Fortunately, some of the ideas used to understand linear systems can be modified to apply to nonlinear systems. In particular, state (or phase) spaces and equilibrium solutions (as well as eigenvalues and eigenvectors) continue to play a key role in understanding the long-term behavior of solutions. You will also see some new phenomena that occur only in nonlinear systems. We restrict our attention to *autonomous* equations, that is, equations in which time does not explicitly appear in the rate functions.

Key words   Nonlinear systems of differential equations; linearization; direction fields; state (phase) space; equilibrium points; Jacobian matrices; eigenvalues; separatrices; bifurcations; limit cycles; predator-prey; van der Pol system; saxophone; spinning bodies; conservative systems; integrals; angular velocity; nonlinear double pendulum

See also   Chapter 6 for background on linear systems and Chapters 8–10 and 12 for more examples of nonlinear systems.

◆ Linear vs. Nonlinear

In modeling a dynamical process with ODEs we aim for a model that is both reasonably accurate and solvable. By the latter we mean that there are either explicit solution formulas that reveal how solutions behave, or reliable numerical solvers for approximating solutions. Constant-coefficient linear ODEs and linear systems have explicit solution formulas (see Chapters 4 and 6), and that is one reason linearity is widely assumed in modeling. However, nonlinearity is an essential feature of many dynamical processes, but explicit solution formulas for nonlinear ODEs are rare. So for nonlinear systems we turn to the alternative approaches, and that's what this chapter is about.

◆ The Geometry of Nonlinear Systems

Let's start with the *linear* system of ODEs that models the motion of a certain viscously damped spring-mass system that obeys Hooke's Law for the displacement x of a unit mass from equilibrium:

$$x' = y, \quad y' = -x - 0.1y \tag{1}$$

In Chapter 4 we saw that the equivalent linear second-order ODE, $x'' + 0.1x' + x = 0$, has an explicit solution formula, which we can use to determine the behavior of solutions and of trajectories in the xy-phase plane.

Now let's suppose that the Hooke's-law spring is replaced by a stiffening spring, which can be modeled by replacing the Hooke's-law restoring force $-x$ in system (1) with the nonlinear restoring force $-x - x^3$. We obtain the system

$$x' = y, \quad y' = -x - x^3 - 0.1y \tag{2}$$

As in the linear system (1), the nonlinear system (2) defines a *vector (or direction) field* in the *xy-state (or phase) plane*. The field lines are tangent to the *trajectories (or orbits)* and point in the direction of increasing time.

There are no solution formulas for system (2), so we turn to direction fields and ODE Architect for visual clues to solution behavior. As you can see from Figure 7.1, the graphs generated by ODE Architect tell us that the trajectories of both systems spiral into the *equilibrium point* at the origin as $t \to +\infty$, even though the shapes of the trajectories differ. The origin corresponds to the constant solution $x = 0$, $y = 0$, which is called a *spiral sink* for each system because of the spiraling nature of the trajectories and because the trajectories, like water in a draining sink, are "pulled" into the origin with the advance of time. This is an indication of *long-term* or *asymptotic* behavior. Note that in this case the nonlinearity does not affect long-term behavior, but clearly does affect short-term behavior.

☞ The equilibrium points of a system correspond to the constant solutions, that is, to the points where all the rate functions of the system are zero.

✓ "Check" your understanding by answering these questions: Do the systems (1) and (2) have any equilibrium points other than the origin? How do the corresponding springs and masses behave as time increases? Why does the $-x^3$ term seem to push orbits toward the y-axis if $|x| \geq 1$, but not have much effect if $|x|$ is close to zero?

◆ Linearization

If we start with a nonlinear system such as (2), we can often use *linear approximations* to help us understand some features of its solutions. Our approximations will give us a corresponding linear system and we can apply what we know about that linear system to try to understand the nonlinear system. In particular, we will be able to verify our earlier conclusions about the long-term behavior of the nonlinear spring-mass system (2).

The nonlinearity of system (2) comes from the $-x^3$ term in the rate function $g(x, y) = -x - x^3 - 0.1y$. In calculus you may have seen the following formula for the linear approximation of the function $g(x, y)$ near the point (x_0, y_0):

☞ This is a *finite Taylor series* approximation to $g(x, y)$.

$$g(x, y) \approx g(x_0, y_0) + \frac{\partial g}{\partial x}(x_0, y_0)(x - x_0) + \frac{\partial g}{\partial y}(x_0, y_0)(y - y_0) \quad (3)$$

However, $g(x_0, y_0)$ will always be zero at an equilibrium point (do you see why?), so formula (3) simplifies in this case to

$$g(x, y) \approx \frac{\partial g}{\partial x}(x_0, y_0)(x - x_0) + \frac{\partial g}{\partial y}(x_0, y_0)(y - y_0) \quad (4)$$

Since we're interested in long-term behavior and the trajectories of system (2) seem to be heading toward the origin, we want to use the equilibrium point

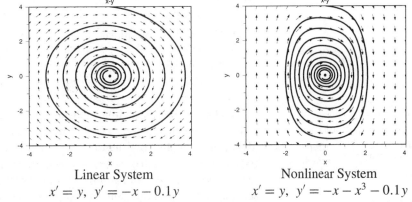

Linear System Nonlinear System
$x' = y, \ y' = -x - 0.1y$ $x' = y, \ y' = -x - x^3 - 0.1y$

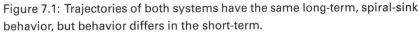

Figure 7.1: Trajectories of both systems have the same long-term, spiral-sink behavior, but behavior differs in the short-term.

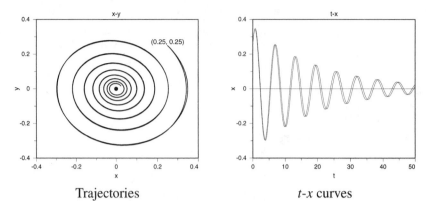

Trajectories   *t-x* curves

Figure 7.2: Near the equilibrium at the origin trajectories and *tx*-component curves of nonlinear system (2) and its linearization (1) are nearly look-alikes.

$(x_0, y_0) = (0, 0)$ in formula (4). Near the origin, the rate function for our nonlinear spring can be approximated by

$$g(x, y) \approx -x - 0.1y$$

since $\partial g / \partial x = -1$ and $\partial g / \partial y = -0.1$ at $x_0 = 0$, $y_0 = 0$. Therefore the nonlinear system (2) reduces to the *linearized system* (1). You can see the approximation when the phase portraits are overlaid. The trajectories and *tx*-component curves of both systems, issuing from a common initial point close to the origin, are shown in Figure 7.2. The linear approximation is pretty good because the nonlinearity $-x^3$ is small near $x = 0$. Take another look at Figure 7.1; the linear approximation is *not* very good when $|x| > 1$.

☞ Linear and nonlinear look-alikes.

✓ How good an approximation to system (2) is the linearized system (1) if the initial point of a trajectory is far away from the origin? Explain what you mean by "good" and "far away."

In matrix notation, linear system (1) takes the form

$$\begin{bmatrix} x \\ y \end{bmatrix}' = \begin{bmatrix} 0 & 1 \\ -1 & -0.1 \end{bmatrix} \begin{bmatrix} x \\ y \end{bmatrix} \tag{5}$$

so the characteristic equation of the system matrix is $\lambda^2 + 0.1\lambda + 1 = 0$. The matrix has eigenvalues $\lambda = (-0.1 \pm i\sqrt{3.99})/2$, making $(0, 0)$ a spiral sink (due to the negative real part of both eigenvalues). This supports our earlier conclusion that was based on the computer-generated pictures in Figure 7.2. The addition of a nonlinear term to a linear system (in this example, a cubic nonlinearity) does not change the stability of the equilibrium point (a sink in this case) or the spiraling nature of the trajectories (suggested by the complex eigenvalues).

☞ Look back at Chapter 6 for more on complex eigenvalues and spiral sinks.

The linear and nonlinear trajectories and the tx-components shown in Figure 7.2 look pretty much alike. This is often the case for a system

$$\mathbf{x}' = \mathbf{F}(\mathbf{x}) \tag{6}$$

☞ The point $\mathbf{x}_0$ is an equilibrium point of $\mathbf{x}' = \mathbf{F}(\mathbf{x})$ if $\mathbf{F}(\mathbf{x}_0) = 0$.

and its *linearization*

$$\mathbf{x}' = \mathbf{A}(\mathbf{x} - \mathbf{x}_0) \tag{7}$$

at an equilibrium point $\mathbf{x}_0$. Let's assume that the dependent vector variable $\mathbf{x}$ has n components $x_1, \ldots x_n$, that $F_1(\mathbf{x}), \ldots, F_n(\mathbf{x})$ are the components of $\mathbf{F}(\mathbf{x})$, and that these components are at least twice continuously differentiable functions. Then the $n \times n$ constant matrix $\mathbf{A}$ in system (7) is the matrix of the first partial derivatives of the components of $\mathbf{F}(\mathbf{x})$ with respect to the components of $\mathbf{x}$, all evaluated at $\mathbf{x}_0$:

$$\mathbf{A} = \begin{bmatrix} \dfrac{\partial \mathbf{F}_1}{\partial x_1} & \cdots & \dfrac{\partial \mathbf{F}_1}{\partial x_n} \\ \vdots & & \vdots \\ \dfrac{\partial \mathbf{F}_n}{\partial x_1} & \cdots & \dfrac{\partial \mathbf{F}_n}{\partial x_n} \end{bmatrix}_{\mathbf{x}=\mathbf{x}_0}$$

$\mathbf{A}$ is called the *Jacobian matrix* of $\mathbf{F}$ at $\mathbf{x}_0$, and is often denoted by $\mathbf{J}$ or $\mathbf{J}(\mathbf{x}_0)$. As an example, look back at system (1) and its linearization, system (2) or system (5).

☞ Here's why linearization is so widely used.

It is known that if none of the eigenvalues of the Jacobian matrix at an equilibrium point is zero or pure imaginary, then close to the equilibrium point the trajectories and component curves of systems (6) and (7) look alike. We can use ODE Architect to find equilibrium points, calculate Jacobian matrices and their eigenvalues, and so, check out whether the eigenvalues meet the conditions just stated. If $n = 2$, we can apply the vocabulary of planar linear systems from Chapter 6 to nonlinear systems. We can talk about a *spiral sink*, a *nodal source*, a *saddle point*, etc. ODE Architect uses a solid dot for a sink, an open dot for a source, a plus sign for a center, and an open square for a saddle.

What happens when, say, the matrix $\mathbf{A}$ *does* have pure imaginary eigenvalues? Then all bets are off, as the following example shows.

Start with the linear system

$$x' = y$$
$$y' = -x$$

The system matrix has the pure imaginary eigenvalues $\pm i$, making the origin a center. Now give the system a nonlinear perturbation to get

$$x' = y - x^3$$
$$y' = -x$$

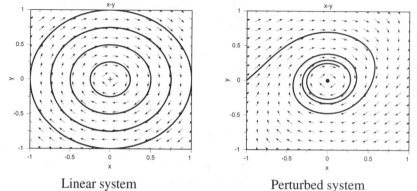

Linear system   Perturbed system

Figure 7.3: Nonlinear terms convert a linear center to a nonlinear sink.

☞ But linearity can be misleading!

By picturing the direction field defined by this system, we can see that each vector has been nudged slightly inward, toward the origin. This causes solutions to spiral inward, making $(0, 0)$ a spiral sink. Figure 7.3 shows trajectories from the original linear system on the left, and a trajectory of the nonlinear system on the right, spiraling inward. Now it should be clear why we had to exclude pure imaginary eigenvalues!

✓   What happens if you perturb the linear system by adding the x^3 term, instead of subtracting? What about the system $x' = y - x^3, \ y' = -x + y^3$?

◆ Separatrices and Saddle Points

A linear saddle point has two trajectories that leave the point (as time increases from $-\infty$) along a straight line in the direction of an eigenvector. Another two trajectories approach the point (as time increases to $+\infty$) along a straight line in the direction of an eigenvector. These four trajectories are called *saddle separatrices* because they divide the neighborhood of the saddle point into regions of quite different long-term trajectory behavior. The left plot in Figure 7.4 shows the four separatrices along the x- and y- axes for the linear system

$$x' = x, \quad y' = -y \tag{8}$$

with a saddle point at the origin. The two that leave the origin as t increases are the *unstable separatrices*, and the two that enter the origin are the *stable separatrices*.

    If we add some higher-order nonlinear terms to a linear saddle-point system, the separatrices persist but their shapes may change. They still divide a neighborhood of the equilibrium point into regions of differing long-term behavior. And, most importantly, they still leave or approach the equilibrium

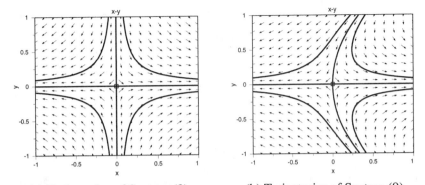

(a) Trajectories of System (8)   (b) Trajectories of System (9)

Figure 7.4: Saddle separatrices lie along the axes in (a); two of the separatrices are bent to the right by a nonlinearity in (b).

point tangent to eigenvectors of the linearized system. The right plot in Figure 7.4 suggests all this for the system (8) with a nonlinear term tacked on:

$$x' = x - y^2, \quad y' = -y \tag{9}$$

Note how the nonlinearity bends two of the separatrices.

◆ Behavior of Solutions Away from Equilibrium Points

While we can use linearization in most cases to determine the long-term behavior of solutions near an equilibrium point, it may not be a good method for studying the behavior of solutions "far away" from the equilibrium point. Consider, for example, the spider-fly system of Module 7:

$$S' = -4S + 2SF, \quad F' = 3\left(1 - \frac{F}{5}\right)F - 2SF$$

where S is a population of spiders preying on F, a population of flies (all measured in thousands). This nonlinear system has several equilibrium points, one of which is at $p^* = (0.9, 2)$.

Take a look at the graphics windows in Experiment 2 of "The Spider and Fly" (Screen 1.5). The trajectories of the linearized system that are close to p^* approximate well those of the nonlinear system. However, trajectories of the linearized system that are not near the equilibrium point diverge substantially from those of the nonlinear system, and may even venture into a region of the state space where the population of spiders is negative!

✓ Look at the Library file "Mutualism: Symbiotic Interactions" in the "Population Models" folder and investigate the long-term behavior of solution curves by using linear approximations near equilibrium points.

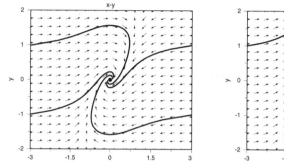
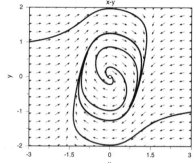

An attracting spiral sink ($a = -1$).   A repelling spiral source and attract-
                                          ing limit cycle ($a = 1$).

Figure 7.5: The system, $x' = y + ax - x^3$, $y' = -x$, undergoes a Hopf bifurcation
to an attracting limit cycle as the parameter transits the value $a = 0$.

◆ Bifurcation to a Limit Cycle

☞ If you are not "into" nonlinear electrical circuits, ignore the modeling here and just consider ODE (10) as a particular nonlinear system.

The model equations for an electrical circuit (the *van der Pol circuit*) containing a nonlinear resistor, an inductor, and a capacitor, all in series, are

$$x' = y + ax - x^3, \quad y' = -x \tag{10}$$

☞ Current, voltages and time are scaled to dimensionless quantities in system (10).

where x is the current in the circuit and y is the voltage drop across the capacitor. The voltage drop across the nonlinear resistor is $ax - x^3$, where a is a
parameter. The characteristics of the resistor, and thus the performance of the
circuit, change when we change the value of this parameter. Let's look at the
phase portrait and the corresponding eigenvalues of the linearization of this
system at the equilibrium point $(0, 0)$ for different values of a.

As a increases from -1 to 1, the eigenvalues of the Jacobian matrix of
system (10) at the origin change from complex numbers with negative real
parts to complex numbers with positive real parts, but at $a = 0$ they are pure
imaginary. The circuit's behavior changes as a increases, and it changes in a
qualitative way at $a = 0$. The phase portrait shows a spiral sink at $(0, 0)$ for
$a \le 0$, then a spiral source for $a > 0$. Further, the trajectories near the source
spiral out to a closed curve that is itself a solution. Our electrical circuit
has gone from one where current and voltage die out to one that achieves a
continuing oscillation described by a periodic steady state. A change like this
in the behavior of a model at a particular value of a parameter is called a *Hopf
bifurcation*. Figure 7.5 shows the changes in a trajectory of system (10) due
to the bifurcation that occurs when a is increased through zero.

✓ Find the Jacobian matrix of system (10) at the origin and calculate its
eigenvalues in terms of the parameter a. Write out the linearized version of
system (10). Check your work by using ODE Architect's equilibrium, Jacobian, and eigenvalue capabilities.

The closed solution curve in Figure 7.5 that represents a periodic steady state is called an *attracting limit cycle* because all nearby trajectories spiral into it as time increases. As the parameter value changes in a Hopf bifurcation, you can observe an equilibrium point that is a spiral sink changing into a source with nearby orbits spiralling onto the limit cycle. You'll investigate this kind of phenomenon when you use ODE Architect to investigate the model system in the "Saxophone" submodule of Module 7.

☞ A limit cycle is exclusively a nonlinear phenomenon. Any cycle in a linear autonomous system is always part of a family of cycles, none of which are limit cycles.

◆ Higher Dimensions

So far we have looked at systems of nonlinear ODEs involving only two state variables. However it is not uncommon for a model to have a system with more than two state variables. Fortunately our ideas extend in a natural way to cover these situations. Analysis by linear approximation may still work in these cases, and ODE Architect can always be used to find equilibrium points, Jacobian matrices, and eigenvalues in any dimension. See for example Problem 4 in Exploration 7.3.

The chapter cover figure shows trajectories of a system with three state variables; this system describes the angular velocity of a spinning body. The "Spinning Bodies" submodule of Module 7 and Problem 1 in Exploration 7.3 model the rotational motion of an object thrown into space; this model is described below.

✓ How could you visualize the trajectories of a system of four equations?

◆ Spinning Bodies: Stability of Steady Rotations

Suppose that a rigid body is undergoing a steady rotation about an axis $\mathbf{L}$ through its center of mass. In a plane perpendicular to $\mathbf{L}$ let θ be the angle swept out by a point in the body, but not on the axis. Steady rotations about $\mathbf{L}$ are characterized by the fact that $\theta' = d\theta/dt = $ constant, for all time. In mechanics, it is useful to describe such steady rotations by a vector $\boldsymbol{\omega}$ parallel to $\mathbf{L}$ whose magnitude $|\boldsymbol{\omega}| = d\theta/dt$ is constant. Notice that $-\boldsymbol{\omega}$ in this case also corresponds to a steady spin about $\mathbf{L}$, but in the opposite direction. The vector $\boldsymbol{\omega}$ is called the *angular velocity*, and for steady rotations we see that $\boldsymbol{\omega}$ is a constant vector. The angular velocity vector $\boldsymbol{\omega}$ can also be defined for an unsteady rotation of the body, but in this case $\boldsymbol{\omega}(t)$ is not a constant vector.

☞ As we shall see, not all axes $\mathbf{L}$ will support steady rotations.

It turns out that in a uniform force field (such as the gravitational field near the earth's surface), the differential equations for the rotational motion of the body about its center of mass decouple from the ODEs for the translational motion of the center of mass. How shall we track the rotational motion of the body? For each rigid body there is a natural triple of orthogonal axes $\mathbf{L}_1$, $\mathbf{L}_2$, and $\mathbf{L}_3$ (called *body axes*) which, as it turns out, makes it relatively easy to model the rotational motion by a system of ODEs. To define the body axes we need the *inertia tensor* **I** of the body. Given a triple of orthogonal axes through the body's center of mass, put an orientation on each axis and label them to form a *right-handed frame* (i.e., it follow the right-hand rule). In that frame, **I** is represented by a 3×3 *positive definite matrix*. Body axes are just the frame for which the representation of **I** is a diagonal matrix with the positive entries I_1, I_2, and I_3 along the diagonal. These values I_1, I_2, and I_3 are called the *principal moments of inertia* of the body. Note that I_k is the moment of inertia about the principal axis $\mathbf{L}_k$, for $k = 1, 2, 3$. If a body has uniform density and an axis **L** such that turning the body $180°$ about that axis brings the body into coincidence with itself again, then that axis **L** is a principal axis.

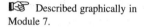 A matrix A is positive definite if it is symmetric (i.e., $A^T = A$) and all of its eigenvalues are positive.

Let's say that a book has uniform density (not quite true, but nearly so). Then the three axes of rotational symmetry through the center of mass are the principal axes: $\mathbf{L}_3$, the *short axis* through the center of the book's front and back covers; $\mathbf{L}_2$, the *long axis* parallel to the book's spine; and $\mathbf{L}_1$, the *intermediate axis* which is perpendicular to $\mathbf{L}_2$ and $\mathbf{L}_3$. For a tennis racket, the body axis $\mathbf{L}_2$ is obvious on geometrical grounds. The other axes $\mathbf{L}_1$ and $\mathbf{L}_3$ are a bit more difficult to discern, but they are given in the margin sketch.

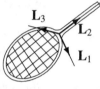 Described graphically in Module 7.

Throw a tennis racket up into the air and watch its gyrations. Wrap a rubber band around a book, toss it into the air, and look at its spinning behavior. Now try to get the racket or the book to spin steadily about each of three perpendicular body axes $\mathbf{L}_1$, $\mathbf{L}_2$, and $\mathbf{L}_3$. Not so hard to do about two of the axes—but nearly impossible about the third. Why is that? Let's construct a model for the rotation of the body and answer this question.

Let's confine our attention to the body's angular motion while aloft, not its vertical motion. Let's ignore air resistance. The key parameters that influence the angular motion are the principal inertias I_1, I_2, I_3 about the respective body axes $\mathbf{L}_1$, $\mathbf{L}_2$, $\mathbf{L}_3$. Let ω_1, ω_2, and ω_3 be the components of the vector ω along the body axes $\mathbf{L}_1$, $\mathbf{L}_2$, and $\mathbf{L}_3$. There is an analogue of Newton's second law applied to the body which involves the angular velocity vector ω. The components of the rotational equation of motion in the body axes frame are given by $I_1\omega_1' = (I_2 - I_3)\omega_2\omega_3$, $I_2\omega_2' = (I_3 - I_1)\omega_1\omega_3$, $I_3\omega_3' = (I_1 - I_2)\omega_1\omega_2$.

A complete derivation of the model ODEs can be found in the first of the listed references.

Dividing by the principal inertias, we have the nonlinear system

$$\omega_1' = \frac{I_2 - I_3}{I_1}\omega_2\omega_3$$

$$\omega_2' = \frac{I_3 - I_1}{I_2}\omega_1\omega_3 \qquad (11)$$

$$\omega_3' = \frac{I_1 - I_2}{I_3}\omega_1\omega_2$$

Let's measure angles in radians and time in seconds, so that each ω_i has units of radians per second.

First, we note that for any constant $\alpha \neq 0$, the equilibrium point $\omega = (\alpha, 0, 0)$ of system (11) represents a pure steady rotation (or spinning motion)

Pure steady rotations are possible about any body axis.

about the first body axis L_1 with angular velocity α. The equilibrium point $(-\alpha, 0, 0)$ represents steady rotation about L_1 in the opposite direction. Similar statements are true for the equilibrium points $\omega = (0, \alpha, 0)$ and $(0, 0, \alpha)$.

Now the *kinetic energy of angular rotation* is given by

$$KE(\omega_1, \omega_2, \omega_3) = \frac{1}{2}\left(I_1\omega_1^2 + I_2\omega_2^2 + I_3\omega_3^2\right)$$

The value of KE stays fixed on an orbit of system (11) since

$$\frac{d(KE)}{dt} = I_1\omega_1\omega_1' + I_2\omega_2\omega_2' + I_3\omega_3\omega_3'$$

$$= (I_2 - I_3)\omega_1\omega_2\omega_3 + (I_3 - I_1)\omega_1\omega_2\omega_3 + (I_1 - I_2)\omega_1\omega_2\omega_3 = 0$$

☞ A system of autonomous ODEs is *conservative* if there is a function F of the dependent variables whose value is constant along each orbit (i.e., trajectory), but varies from one orbit to another. F is said to be an *integral of motion* of the system.

So system (11) is *conservative* and KE is an *integral*. The ellipsoidal integral surface $KE = C$, where C is a positive constant, is called an *inertial ellipsoid* for system (11). Note that any orbit of (11) that starts on one of the ellipsoids stays on the ellipsoid, and orbits on that ellipsoid share the same value of KE.

✓ Show that the functions

$$K = \frac{I_3 - I_1}{I_2}\omega_1^2 - \frac{I_2 - I_3}{I_1}\omega_2^2$$

and

$$M = \frac{I_1 - I_2}{I_3}\omega_1^2 - \frac{I_2 - I_3}{I_1}\omega_3^2$$

are also integrals for system (11). Describe the surfaces $K = $ const., $M = $ const.

Let's put in some numbers for I_1, I_2, and I_3 and see what happens. Set $I_1 = 2$, $I_2 = 1$, $I_3 = 3$. Then system (11) becomes

$$\omega_1' = -\omega_2\omega_3$$
$$\omega_2' = \omega_1\omega_3 \tag{12}$$
$$\omega_3' = \frac{1}{3}\omega_1\omega_2$$

With the given values for I_1, I_2, I_3 we have the integral

$$KE = \frac{1}{2}\left(2\omega_1^2 + \omega_2^2 + 3\omega_3^2\right) \tag{13}$$

☞ Body axis L_1 is parallel to the ω_1-axis, L_2 to the ω_2-axis, and L_3 to the ω_3-axis in Figure 7.6.

The left graph in Figure 7.6, which is also the chapter cover figure, shows the inertial ellipsoid $KE = 12$ and twenty-four orbits on the surface. The geometry of the orbits indicates that if the body is started spinning about an axis very near the body axes L_2 or L_3, then the body continues to spin almost steadily about those body axes. Attempting to spin the body about the intermediate body axis L_1 is another matter. Any attempt to spin the body about the L_1 body axis leads to strange gyrations. Note in Figure 7.6 that each of the four trajectories that starts near the equilibrium point $(\sqrt{12}, 0, 0)$ where the

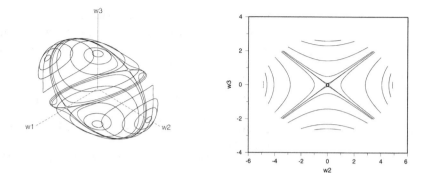

Figure 7.6: Twenty-four trajectories on the inertial ellipsoid $KE = 12$ (left); head-on view from the ω_1-axis (right) shows a saddle point on the ellipsoid.

ω_1-axis pierces the ellipsoid goes back near the antipodal point (and reverses its direction of rotation) then returns in an endlessly repeating periodic path. This corresponds to unstable gyrations near the ω_1-axis.

✓   Match up the trajectories in Figure 7.6 with actual book rotations. Put a rubber band around a book, flip the book into the air, and check out the rotations. Do the projected trajectories in the right graph of Figure 7.6 really terminate, or is something else going on?

◆ The Planar Double Pendulum

☞ This is pretty advanced material here, so feel free to skip the text and go directly to the "Double Pendulum Movies". Just click on the ODE Architect library, open the "Physical Models" folder and the "Double Pendulum Animator" file, and create chaos!

The planar double pendulum is an interesting physical system with two degrees of freedom. It consists of two rods of lengths l_1 and l_2, and two masses, specified by m_1 and m_2, attached together so that the rods are constrained to oscillate in a vertical plane. We'll neglect effects of damping in this system.

The governing equations are most conveniently written in terms of the angles $\theta_1(t)$ and $\theta_2(t)$ shown in Figure 7.7. One way to obtain the equations of motion is by applying Newton's law to the motions of the masses. First we'll consider mass m_2 and the component in the direction shown by the unit vector $\mathbf{u}_3$ in Figure 7.7. Define a coordinate system centered at mass m_1 and rotating with angular velocity $\boldsymbol{\Omega} = (d\theta_1/dt)\mathbf{k}$, where $\mathbf{k}$ is the unit vector normal to the plane of motion. If $\hat{\mathbf{a}}$, $\hat{\mathbf{v}}$, and $\hat{\mathbf{r}}$ denote the acceleration, velocity, and position of m_2 with respect to the rotating coordinate system, then the acceleration $\mathbf{a}$ with respect to a coordinate system at rest is known to be

$$\mathbf{a} = \hat{\mathbf{a}} + \frac{d\boldsymbol{\Omega}}{dt} \times \hat{\mathbf{r}} + 2\boldsymbol{\Omega} \times \hat{\mathbf{v}} + \boldsymbol{\Omega} \times (\boldsymbol{\Omega} \times \hat{\mathbf{r}}) \tag{14}$$

For our configuration it follows that

$$\hat{\mathbf{r}} = -[l_1 \sin(\theta_2 - \theta_1)]\mathbf{u}_3 + [l_2 + l_1 \cos(\theta_2 - \theta_1)]\mathbf{u}_4 \tag{15}$$

with the unit vector $\mathbf{u}_4$ in the direction shown in Figure 7.7. Since the only forces acting are gravity and the tensile forces in the rods, the $\mathbf{u}_3$-component of $\mathbf{F} = m_2\mathbf{a}$ in combination with Eqs. (14) and (15) gives

$$m_2 l_2 (\theta_2 - \theta_1)'' + m_2 [l_2 + l_1 \cos(\theta_2 - \theta_1)] \theta_1''$$
$$+ m_2 l_1 (\theta_1')^2 \sin(\theta_2 - \theta_1) = -m_2 g \sin\theta_2 \quad (16)$$

Similarly, the component of Newton's law in the direction of the unit vector $\mathbf{u}_1$ is given by

$$m_2 l_1 \theta_1'' + m_2 l_2 \cos(\theta_2 - \theta_1)\theta_2'' - m_2 l_2 (\theta_2')^2 \sin(\theta_2 - \theta_1)$$
$$= -m_2 g \sin\theta_1 - f_2 \sin(\theta_2 - \theta_1) \quad (17)$$

where f_2 is the magnitude of the tensile force in the rod l_2. Equations (16) and (17) will provide the system governing the motion, once the quantity f_2 is determined. An equation for f_2 is found from the $\mathbf{u}_1$-component of Newton's law applied to the mass m_1:

$$m_1 l_1 \theta_1'' = -m_1 g \sin\theta_1 + f_2 \sin(\theta_2 - \theta_1) \quad (18)$$

Eliminating f_2 between Eqs. (17) and (18) and simplifying Eq. (16) slightly, we obtain the governing nonlinear system of second-order ODEs for the planar double pendulum:

$$(m_1 + m_2) l_1 \theta_1'' + m_2 l_2 \cos(\theta_2 - \theta_1)\theta_2''$$
$$- m_2 l_2 (\theta_2')^2 \sin(\theta_2 - \theta_1) + (m_1 + m_2) g \sin\theta_1 = 0 \quad (19)$$

$$m_2 l_2 \theta_2'' + m_2 l_1 \cos(\theta_2 - \theta_1)\theta_1''$$
$$+ m_2 l_1 (\theta_1')^2 \sin(\theta_2 - \theta_1) + m_2 g \sin\theta_2 = 0 \quad (20)$$

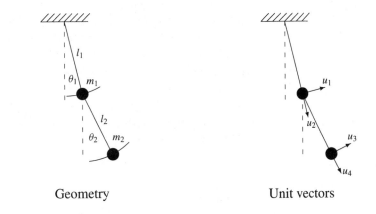

Geometry   Unit vectors

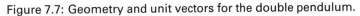

Figure 7.7: Geometry and unit vectors for the double pendulum.

Another way to derive the equations of motion of the double pendulum system is to use Lagrange's equations. These are

$$\frac{d}{dt}\left[\frac{\partial}{\partial\theta_1'}(T-V)\right]-\frac{\partial}{\partial\theta_1}(T-V)=0 \tag{21}$$

$$\frac{d}{dt}\left[\frac{\partial}{\partial\theta_2'}(T-V)\right]-\frac{\partial}{\partial\theta_2}(T-V)=0 \tag{22}$$

where T is the kinetic energy of the system and V is its potential energy. The respective kinetic energies of the masses m_1 and m_2 are

$$T_1=\frac{1}{2}m_1 l_1^2(\theta_1')^2$$

$$T_2=\frac{1}{2}m_2(l_1\theta_1'\sin\theta_1+l_2\theta_2'\sin\theta_2)^2+\frac{1}{2}m_2(l_1\theta_1'\cos\theta_1+l_2\theta_2'\cos\theta_2)^2$$

The corresponding potential energies of m_1 and m_2 are

$$V_1=m_1 g l_1(1-\cos\theta_1)$$

$$V_2=m_2 g l_1(1-\cos\theta_1)+m_2 g l_2(1-\cos\theta_2)$$

Then we have $T=T_1+T_2$ and $V=V_1+V_2$. Inserting the expressions for T and V into Eqs. (21) and (22), we find the equations of motion of the double pendulum. These equations are equivalent to the ones obtained previously using Newton's law. The formalism of Lagrange pays the dividend of producing the equations with "relatively" shorter calculations.

References   Borrelli, R.L., Coleman, C.S., and Boyce, W.E., *Differential Equations Laboratory Workbook*, (1992: John Wiley & Sons, Inc.), Experiment 6.9

Hirsch, M., and Smale, S., *Differential Equations, Dynamical Systems, and Linear Algebra* (1974: Academic Press), Chapter 10

Hubbard, J., and West, B., *Differential Equations: A Dynamical Systems Approach*, Vol. 18 of Texts in Appl. Math. (1995: Springer-Verlag), Sections 8.2, 8.3

Marion, J.B., and Thornton, S.T., *Classical Dynamics of Particles and Systems*, 3rd ed. (1988: Harcourt Brace Jovanovich)

Answer questions in the space provided, or on attached sheets with carefully labeled graphs. A notepad report using the Architect is OK, too.

Name/Date _____

Course/Section _____

Exploration 7.1. Predator and Prey: Linearization and Stability

1. Let F represent the number of flies and S the number of spiders (both in 1000s). Assume that the model for their interaction is given by:

    ☞ Take a look at the "Spider and Fly" submodule of Module 7.

    $$S' = -4S + 2SF, \quad F' = 3F - 2SF \tag{23}$$

    where the SF-term is a measure of the interaction between the two species.

    (a) Why is the SF-term negative in the first ODE and positive in the second when (S, F) is inside the population quadrant?

    (b) Show that the system has an equilibrium point at $(1.5, 2)$.

    (c) Show that the system matrix of the linearization of system (23) about $(1.5, 2)$ has pure imaginary eigenvalues.

    ☞ This makes the point $(1.5, 2)$ a center for the linearized system.

    (d) Now plot phase portraits for system (23) and for its linearization about $(1.5, 2)$. What do you see?

2. Suppose that an insecticide reduces the spider population at a rate proportional to the size of the population.

    (a) Modify the predator-prey model of system (23) to account for this.

    (b) Model how insecticide can be made more or less effective.

    (c) Use the model to predict the long-term behavior of the populations.

3. In a predator-prey system that models spider-fly interaction

$$S' = -4S + 2SF, \quad F' = 3\left(1 - \frac{F}{N}\right)F - 2SF$$

the number N represents the maximum fly population (in 1000s). Investigate the effect of changing the value of N. What's the largest the spider population can get? The fly population?

4. Suppose the spider-fly model is modified so that there are two predators, spiders and lizards, competing to eat the flies. One model for just the two predator populations is

$$S' = 4\left(1 - \frac{S}{5}\right)S - SL, \quad L' = 3\left(1 - \frac{L}{2}\right)L - SL$$

   (a) What do the numbers 2, 3, 4, and 5 represent?
   (b) What does the term SL represent? Why is it negative?
   (c) What will become of the predator populations in the long run?

5. Take a look at the library file "A Predator-Prey System with Resource Limitation" in the "Biological Models" folder. Compare and contrast the system you see in that file with that given in Problem 2. Create a system where both the predator and the prey are subject to resource limitations, and analyze the behavior of the trajectories.

Answer questions in the space provided, or on
attached sheets with carefully labeled graphs. A
notepad report using the Architect is OK, too.

Name/Date _____

Course/Section _____

Exploration 7.2. Bifurcations and Limit Cycles

1. Alter the model in the "Saxophone" submodule of Module 10 by adding a
    parameter c:

    $$u' = v, \quad v' = -su + cv - \frac{1}{b}v^3$$

    (a) What part of the model does this affect?

    (b) How do solutions behave for values of c between 0 and 2, taking $s = b = 1$?

    (c) As c increases, what happens to the pitch and amplitude?

2. Suppose the model for a *simple harmonic oscillator* (a linear model),

    $$x' = y, \quad y' = -x$$

    is modified by adding a parameter c:

    $$x' = cx + y, \quad y' = -x + cy$$

    (a) What happens to the equilibrium point as c goes from -1 to 1?

    (b) What happens to the eigenvalues of the matrix of coefficients as c changes
        from -1 to 1?

3. Suppose we further modify the system of Problem 2:

$$x' = cx + y - x(x^2 + y^2), \qquad y' = -x + cy - y(x^2 + y^2)$$

where $-1 \leq c \leq 1$. Analyze the behavior of the equilibrium point at $(0, 0)$ as c increases from -1 to 1. How does it compare with the behavior you observed in Problem 2?

4. You can modify the system for a simple, undamped nonlinear pendulum (see Chapter 10) to produce a *torqued pendulum*:

$$x' = y, \qquad y' = -\sin(x) + a$$

☞ Use the model-based pendulum animation in ODE Architect and watch the pendulum gyrate.

Here a represents a torque applied about the axis of rotation of the pendulum arm. Investigate the behavior of this torqued pendulum for the values of a between 0 and 2 by building the model and animating the phase space as a increases. Explain what kind of behavior the pendulum exhibits as a increases; explain the behavior of any equilibrium points you see.

5. The motion of a thin, flexible steel beam, affixed to a rigid support over two magnets, can be modeled by *Duffing's equation*:

$$x' = y, \quad y' = ax - x^3$$

☞ Sweep on the parameter a, and then animate. To animate a graph with multiple trajectories corresponding to different values of a, click on the animate icon below the word "Tools" at the top left of the tools screen.

where x represents the horizontal displacement of the beam from the rest position and a is a parameter that is related to the strength of the magnets. Investigate the behavior of this model for $-1 \leq a \leq 1$. In particular:

(a) Find all equilibrium points and classify them as to type (e.g., center, saddle point), verifying your phase plots with eigenvalue calculations (use ODE Architect for the eigenvalue calculations). Some of your answers will depend on a.

(b) Give a physical interpretation of your answers to Question (a).

(c) What happens to the equilibrium points as the magnets change from weak ($a \leq 0$) to strong ($a > 0$)?

(d) What happens if you add a linear damping term to the model? (Say, $y' = ax - x^3 - vy$.)

Answer questions in the space provided, or on attached sheets with carefully labeled graphs. A notepad report using the Architect is OK, too.

Name/Date _____

Course/Section _____

Exploration 7.3. Higher Dimensions

1. *Spinning Bodies.*
   Use ODE Architect to draw several distinct trajectories on the ellipsoid of inertia, $0.5(2\omega_1^2 + \omega_2^2 + 3\omega_3^2) = 6$, for system (12).

   ☞ For example, $\omega_1 = 0$, $\omega_2 = 3$, $\omega_3 = 1$ can be taken as an initial point and so can $\omega_1 = 1$, $\omega_2 = 0$, $\omega_3 = (10/3)^{0.5}$.

   Choose initial data on the ellipsoid so that the trajectories become the "visible skeleton" of the invisible ellipsoid. What do the trajectories look like? What kind of motion does each represent? You should be able to get a picture that resembles the chapter cover figure and Figure 7.6. Project your 3D graphs onto the $\omega_1\omega_2$-, $\omega_2\omega_3$-, and $\omega_1\omega_3$-planes, and describe what you see. Now apply the equilibrium/eigenvalue/eigenvector calculations from ODE Architect to equilibrium points on each of the ω_1-, ω_2-, and ω_3- axes. Describe the results and their correlation with what you saw on the coordinate planes. Now go to the Library file "A Conservative System: The Momentum Ellipsoid" in the folder "Physical Models" and explain what you see in terms of the previous questions in this problem.

2. Exploration 7.1 (Problem 4) gives a predator-prey model where two species, spiders and lizards, prey on flies. Construct a system of three differential equations that includes the prey in the model. You'll need to represent growth rates and interactions, and you may want to limit population sizes. Make some reasonable assumptions about these parameters. What long-term behavior does your model predict?

3. Take another look at the ODEs of the coupled springs model in Module 6. Use ODE Architect for the system of four ODEs given in Experiment 1 of that section. Make 3D plots of any three of the five variables x_1, x_1', x_2, x_2', and t. What do the plots tell you about the corresponding motions of the springs?

4. Modify the coupled springs model from Module 6 (where coupled linear springs move on a frictionless horizontal surface) by making one of the springs hard or soft: add a term like $\pm x^3$ to the restoring force. Does this change the long-term behavior of the system? Make and interpret graphs as in Problem 3.

5. *Chaos in three dimensions.*

   Some nonlinear 3D systems seem to behave chaotically. Orbits stay bounded as time advances, but the slightest change in the initial data leads to an orbit that eventually seems to be completely uncorrelated with the original orbit. This is thought to be one feature of chaotic dynamics. Choose one of the following three Library files located in the folder "Higher Dimensional Systems":

   - "The Scroll Circuit: Organized Chaos"
   - "The Lorenz System: Chaos and Sensitivity"
   - "The Roessler System: A Strange Attractor"

   Change parameters until you see an example of this kind of chaos. You may want to look at Chapter 12 for additional insight into the meaning of chaos.

8 Compartment Models

Oscillating chemical reactions on a wineglass?

Overview   A salt solution is pumped through a series of tanks. We'll use the *balance law* to model the rate of change of the amount of salt in each tank:

$$\left\{ \begin{array}{c} \text{Net rate of change of} \\ \text{amount of salt in tank} \end{array} \right\} = \left\{ \begin{array}{c} \text{Rate into} \\ \text{tank} \end{array} \right\} - \left\{ \begin{array}{c} \text{Rate out of} \\ \text{tank} \end{array} \right\}$$

If we know the initial amount of salt and the inflow and outflow rates of the solution in each tank, then we can set up an IVP that models the physical system. We'll use this "balance law" approach to model the pollution level in a lake; the flow of a medication; the movement of lead among the blood, tissues, and bones of a body; and an autocatalytic chemical reaction.

Key words   Compartment model; balance law; lake pollution; pharmacokinetics; chemical reactions; chemical law of mass action; autocatalysis; Hopf bifurcation

See also   Chapter 9 for more compartment models, and Chapter 6 for linear systems and flow through interconnected tanks.

◆ Lake Pollution

Modeling how pollutants move through an environment is important in the prediction of harmful effects, as well as the formulation of environmental policies and regulations. The simplest situation has a single source of pollution that contaminates a well-defined habitat, such as a lake. To build a model of this system, we picture the lake as a *compartment*; pollutants in the water flow into and out of the compartment. The rates of flow determine the amount of build-up or dissipation of pollutants. It is useful to represent this conceptual model with a *compartment diagram*, where a box represents a compartment and an arrow represents a flow rate. Here is a compartment diagram for a simple model of lake pollution:

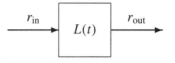

The amount of pollutant in the lake at time t is $L(t)$, while r_{in} is the rate of flow of pollutant into the lake and r_{out} is the rate of flow of pollutant out of the lake. To obtain the equation for the rate of change of the amount of pollutant in the lake, we apply the *balance law*: the net rate of change of the amount of a substance in a compartment is the difference between the rate of flow into the compartment and the rate of flow out of the compartment:

$$\frac{dL}{dt} = r_{\text{in}} - r_{\text{out}}$$

This ODE is sufficient when we know the rates r_{in} and r_{out}, but these rates are usually not constant: they depend on the rate of flow of water into the lake, the rate of flow of water out of the lake, and the pollutant concentration in the inflowing water. Let s_{in} and s_{out} represent the volume rates of flow of water into and out of the lake, V the volume of water in the lake, and p_{in} the concentration of pollutant in the incoming water. Now we can calculate the rates shown in the compartment diagram:

☞ You can get the volume $V(t)$ of water in the lake by solving the IVP $V' = s_{in} - s_{out}$, $V(0) = V_0$.

$$r_{\text{in}} = p_{\text{in}}s_{\text{in}}, \quad r_{\text{out}} = \frac{L}{V}s_{\text{out}}$$

The ODE for the amount of pollutant in the lake is now

$$\frac{dL}{dt} = p_{\text{in}}s_{\text{in}} - \frac{L}{V}s_{\text{out}} \tag{1}$$

☞ So we need to know $V(0)$, $L(0)$, p_{in}, s_{in}, and s_{out} in order to determine $L(t)$ and $V(t)$.

☞ Take a look as Screen 1.4 in Module 8 for on-off inflow concentrations.

To obtain an IVP, we need to specify $L(0)$, the initial amount of pollutant in the lake. The solution to this IVP will reveal how the level of pollution varies in time. Figure 8.1 shows a solution to the ODE (1) for the pollution level in the lake if the inflow is contaminated for the first six months of every year and is clean for the last six months (so $p_{in}(t)$ is a square wave function).

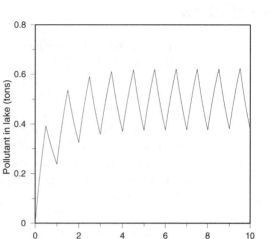

Figure 8.1: Pollutant level in a lake (on-off inflow rates).

✓ "Check" your understanding by finding the volume $V(t)$ of water in the lake at time t if $V(0) = 10$, $s_{in} = 3$, and $s_{out} = 1$, 3, or 5 (all quantities in suitable units). Does the lake dry up, overflow, or stay at a constant volume?

◆ Allergy Relief

Medications that relieve the symptoms of hay fever often contain an antihistamine and a decongestant bundled into a single capsule. The capsule dissolves in the gastrointestinal (or GI) tract and the contents move through the intestinal walls and into the bloodstream at rates proportional to the amounts of each medication in the tract. The kidneys clear medications from the bloodstream at rates proportional to the amounts in the blood.

Here is a compartment diagram for this system:

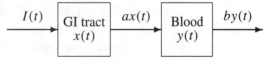

The symbols in this diagram have the following meanings:

$I(t)$: The rate at which the dissolving capsule releases a medication (for example, a decongestant) into the GI tract

$x(t)$: The amount of medication in the GI tract at time t

$ax(t)$: The clearance rate of the medication from the GI tract, which equals the entrance rate into the blood (a is a positive rate constant)

$y(t)$: The amount of medication in the blood at time t

$by(t)$: The clearance rate of the medication from the blood (b is a positive rate constant)

Applying the balance law to each compartment, we have a system of first-order linear ODEs:

$$x' = I - ax$$
$$y' = ax - by \qquad (2)$$

☞ Medication levels in the blood (easily measured) indicate the levels in the tissues (hard to measure), where the medication does its good work.

If you know $I(t)$, the rate constants a and b, and the initial amounts $x(0)$ and $y(0)$ of medication in the GI tract and the bloodstream, you can use ODE Architect to track the flow of the medication through the body. From a pharmacological point of view, the goal is to get the medication levels in the blood into the effective (but safe) zone as quickly as possible and then to keep them there until the patient recovers.

There are two kinds of medication-release mechanisms: continuous and on-off. In the first kind, the medication is released continuously at an approximately constant rate, so $I(t)$ is a positive constant. In the on-off case, each capsule releases the medication at a constant rate over a brief span of time and then the process repeats when the next capsule is taken. In this case we model $I(t)$ by a square wave:

$$I(t) = A \text{ SqWave}(t, T_{per}, T_{on})$$

☞ Screen 2.4 in Module 8 shows what happens if $a = 0.6931 \text{ hr}^{-1}$, $T_{on} = 1$ hr, and b and A are adjustable parameters.

which has amplitude A, period T_{per}, and "on" time T_{on}. For example, if the capsule releases 6 units of medication over a half hour and the dosage is one capsule every six hours, then

$$I(t) = 12 \text{ SqWave}(t, 6, 0.5) \qquad (3)$$

Note that 12 (units/hr)$\times$0.5 (hr) = 6 units.

Compartment models described by equations such as (2) are called *cascades*. They can be solved explicitly, one equation at a time, by solving the first ODE, inserting the solution into the second ODE, solving it, and so on down the cascade. Although this approach theoretically yields explicit solution formulas, in practice the formulas farther along in the cascade of solutions get so complicated that they are difficult to interpret. That's one reason why

☞ ODE Architect to the rescue!

it pays to use a numerical solver, like the ODE Architect. Figure 8.2 shows how the amounts of decongestant in the body change when administered by the on-off method [equation (3)].

✓ By inspecting Figure 8.2 decide which of the clearance coefficients a or b is larger.

References   Borrelli, R.L., and Coleman, C.S., *Differential Equations: A Modeling Perspective*, (1998: John Wiley & Sons, Inc.)

Spitznagel, E., "Two-Compartment Phamacokinetic Models" in *C·ODE·E*, Fall, 1992, pp. 2–4, http://www.math.hmc.edu/codee

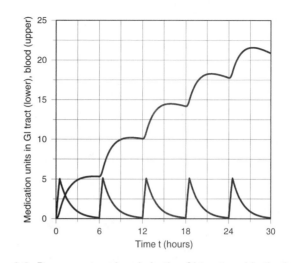

Figure 8.2: Decongestant levels in the GI tract and in the blood.

◆ Lead in the Body

☞ In ancient times lead was used to sweeten wine.

Lead gets into the digestive and respiratory systems of the body via contaminated food, air, and water, as well as lead-based paint, glaze, and crystalware. Lead moves into the bloodstream, which then distributes it to the tissues and bones. From those two body compartments it leaks back into the blood. Lead does the most damage to the brain and nervous system (treated here as tissues). Hair, nails, and perspiration help to clear lead from the tissues, and the kidneys clear lead from the blood. The rate at which lead leaves one compartment and enters another has been experimentally observed to be proportional to the amount that leaves the first compartment. Here is the compartment diagram that illustrates the flow of lead through the body.

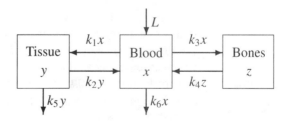

In the diagram, L is the inflow rate of lead into the bloodstream (from the lungs and GI tract); x, y, and z are the respective amounts of lead in the blood, tissues, and bones; and $k_1, \ldots, k_6$ are experimentally determined positive rate constants. The amount of lead is measured in micrograms (1 microgram = 10^{-6} gram), and time (t) is measured in days.

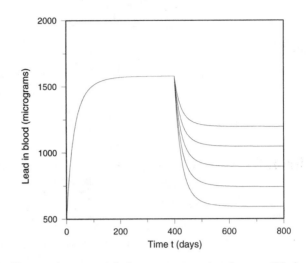

Figure 8.3: Five environmental clean-up scenarios for $t > 400$ days result in five different steady-state lead levels in the blood.

☞ System (4) is a driven linear system with constant coefficients, so eigenvalue/eigenvector techniques can be used to find solution formulas if L is a constant (see Chapter 6).

Applying the balance law to each compartment, we have the linear system of ODEs that models the flow of lead through the body compartments:

$$x' = (L + k_2 y + k_4 z) - (k_1 + k_3 + k_6)x$$
$$y' = k_1 x - (k_2 + k_5) y \tag{4}$$
$$z' = k_3 x - k_4 z$$

Unlike the allergy relief system (2), system (4) is not a cascade. Lead moves back and forth between compartments, so the system cannot be solved one ODE at a time. ODE Architect can be used to find $x(t)$, $y(t)$, and $z(t)$ if $x(0)$, $y(0)$, $z(0)$, $L(t)$, and $k_1, \ldots, k_6$ are known.

If the goal is to reduce the amount of lead in the blood (and therefore in the tissues and bones), we can clean up the environment (which reduces the inflow rate) or administer a medication that increases the clearance coefficient k_6. However, such medication carries its own risks, so most efforts today are aimed at removing lead from the environment. A major step in this direction was made in the 1970's and 80's when oil companies stopped adding lead to gasoline and paint manufacturers began to use other spreading agents in place of lead. Figure 8.3 shows the effects of changing the lead intake rate L.

☞ It was in the 1970's and 80's that most of the environmental protection laws were enacted.

The Food and Drug Administration and the National Institutes of Health have led the fight against lead pollution in the environment. They base their efforts on data acquired from several controlled studies of lead flow, where the study groups were made up of human volunteers in urban areas. The numbers we use in Submodule 3 of Module 8 and in this chapter come from one of those studies. Some references on the lead problem are listed below.

☞ See Screen 3.3 in Module 8 for the rate constants and the inflow rate L.

✓ Write down the systems of ODEs for the two compartment diagrams:

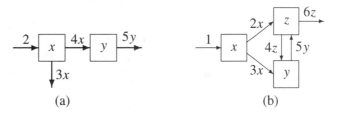

(a)   (b)

References   Batschelet, E., Brand, L., and Steiner, A., "On the kinetics of lead in the human body," *J. Math. Bio.*, **8** (1979), pp. 15–23

Kessel, I., and O'Conner, J., *Getting the Lead Out* (1997: Plenum)

Rabinowitz, M., Wetherill, G., and Kopple, J., "Lead metabolism in the normal human: Stable isotope studies," *Science*, **182** (1973), pp. 725–727.

◆ Equilibrium

In many compartment models, if the inflow rates from outside the system are constant, then the substance levels in each compartment tend to an equilibrium value as time goes on. Mathematically, we can find the equilibrium values by setting each rate equal to zero and solving the resulting system of equations simultaneously. For example, the equilibrium for the system

$$x' = 1 - 2x$$
$$y' = 2x - 3y \tag{5}$$

is $x = 1/2$, $y = 1/3$, which is the solution to the algebraic system $1 - 2x = 0$ and $2x - 3y = 0$. If the system is complicated, you can use ODE Architect to find the equilibrium values. Just use the Equilibrium tabs in the lower left quadrant and in one of the right quadrants, and you will get approximate values for the equilibrium levels.

☞ The Equilibrium tabs in ODE Architect work for systems such as (5), where the rate functions don't depend explicitly on time (i.e., the systems are autonomous).

✓ Go to Things-to-Think-About 2 on Screen 3.5 of Module 8 for the lead flow model with constant values for L and the coefficients k_j. Use the Equilibrium tabs in the tool screen to estimate the equilibrium lead levels in the blood, tissues, and bones for the given data.

☞ You need to know about matrices to tackle this one.

✓ Suppose that x is a column vector with n entries, b is a column vector of n constants, and A is an $n \times n$ invertible matrix of real constants. Can you explain why the linear system $x' = Ax - b$ has a constant equilibrium x^*? Find a formula for x^* in terms of A^{-1} and b.

◆ The Autocatalator and a Hopf Bifurcation

So far all the compartments in our models have represented physical spaces through which substances move. However, there are other ways to think about compartments. For example, they can represent substances that transform into one another, such as uranium 238 and all of its seventeen radioactive decay products, ending with stable lead 206. Or think of a chemical reactor in which chemicals react with one another and produce other chemicals. The *autocatalator* is a mathematical model for one of these chemical reactions.

In an *autocatalytic reaction*, a chemical promotes its own production. For example, suppose that one unit of chemical X reacts with two units of chemical Y to produce three units of Y, a net gain of one unit of Y:

$$X + 2Y \xrightarrow{k} 3Y$$

where k is a positive rate constant. This is an example of *autocatalysis*. We'll come back to autocatalysis, but first we need to make a quick survey of how chemical reactions are modeled by ODEs.

Most chemical reactions are *first-order* in the sense that the rate of decay of each chemical in the reaction is directly proportional to its own concentration:

$$\frac{dz}{dt} = -kz \tag{6}$$

where $z(t)$ is the concentration of chemical Z at time t in the reactor and k is a positive rate constant.

While a first-order reaction is modeled by a *linear* ODE, such as (6), autocatalytic reactions are higher-order and the corresponding rate equations are *nonlinear*. In order to build models of higher-order chemical reactions, we will use a basic principle called the *Chemical Law of Mass Action*:

The Chemical Law of Mass Action. If molecules $X_1, \ldots, X_n$ react to produce molecules $Y_1, \ldots, Y_m$ in one step of the chemical reaction

$$X_1 + \cdots + X_n \xrightarrow{k} Y_1 + \cdots + Y_m$$

that occurs with rate constant k, then

$$x_i' = -kx_1 x_2 \cdots x_n, \ 1 \le i \le n$$
$$y_j' = kx_1 x_2 \cdots x_n, \ 1 \le j \le m$$

where x_i and y_j are, respectively, the concentrations of X_i and Y_j. The chemical species $X_1, \ldots, X_n, Y_1, \ldots, Y_m$ need not be distinct from each other: more than one molecule of a given type may be involved in the reaction.

For example, the chemical law of mass action applied to the reaction

$$X + Y \xrightarrow{k} Z$$

gives

$$x' = -kxy, \quad y' = -kxy, \quad \text{and} \quad z' = kxy$$

where k is a positive rate constant and x, y, z denote the respective concentrations of the chemicals X, Y, Z in the reactor. The autocatalytic reaction

$$X + 2Y \xrightarrow{k} 3Y$$

is modeled by

$$x' = -kxy^2$$
$$y' = -2kxy^2 + 3kxy^2 = kxy^2$$

because the rate of decrease of the reactant concentration x is kxy^2 (think of $X + 2Y$ as $X + Y + Y$), the rate of decrease of the reactant concentration y is $2kxy^2$ (because two units of Y are involved), and the rate of increase in the product concentration y is $3kxy^2$ (think of $3Y$ as $Y + Y + Y$).

✓   If you want to speed up the reaction should you increase the rate constant k, or lower it? Any guesses about what would happen if you heat up the reactor? Put the reactor on ice?

With this background, we can model a sequence of reactions that has been studied in recent years:

$$X_1 \xrightarrow{k_1} X_2, \qquad X_2 \xrightarrow{k_2} X_3, \qquad X_2 + 2X_3 \xrightarrow{k_3} 3X_3, \qquad X_3 \xrightarrow{k_4} X_4$$

Note the nonlinear autocatalytic step in the midst of the first-order reactions. A compartment diagram for this reaction is

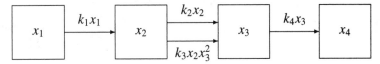

where x_1, x_2, x_3, and x_4 denote the respective concentrations of the chemicals X_1, X_2, X_3, and X_4. The corresponding ODEs are:

☞ The term $k_3x_2x_3^2$ makes this system nonlinear.

$$x_1' = -k_1x_1$$
$$x_2' = k_1x_1 - (k_2x_2 + k_3x_2x_3^2)$$
$$x_3' = (k_2x_2 + k_3x_2x_3^2) - k_4x_3 \tag{7}$$
$$x_4' = k_4x_3$$

☞ See Screen 4.3 of Module 8 for values of the rate constants.

In a reaction like this, we call X_1 the reactant, X_2 and X_3 intermediates, and X_4 the final product of the reaction. For certain ranges of values for the rate constants k_1, k_2, k_3, k_4 and for the initial reactant concentration $x_1(0)$, the

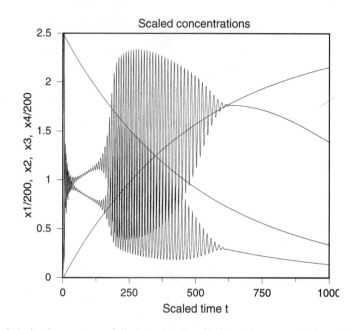

Figure 8.4: As the reactant falls into the Hopf bifurcation zone, the oscillations of the intermediates turn on as the product rises. Later the oscillations turn off and the reaction approaches completion.

☞ The chapter cover figure shows how the intermediate concentrations $x_2(t)$ and $x_3(t)$ play off against each other as time increases.

intermediate concentrations $x_2(t)$ and $x_3(t)$ will suddenly begin to oscillate. These oscillations eventually stop and the intermediates decay into the final reaction product. See Figure 8.4.

The onset of these oscillations is a kind of a *Hopf bifurcation* for $x_2(t)$ and $x_3(t)$. In this context, if we keep the value of x_1 fixed at, say x_1^*, the rate term $k_1 x_1^*$ in system (7) can be viewed as a parameter c. Then the middle two rate equations can be decoupled from the other two:

$$\begin{aligned} x_2' &= c - k_2 x_2 - k_3 x_2 x_3^2 \\ x_3' &= k_2 x_2 + k_3 x_2 x_3^2 - k_4 x_3 \end{aligned} \tag{8}$$

Now let's fix k_2, k_3, and k_4 and use the parameter c to turn the oscillations in $x_2(t)$ and $x_3(t)$ on and off. This is the setting for a *Hopf bifurcation*, so let's take a detour and explain what that is.

As a parameter transits a bifurcation value the behavior of the state variables suddenly changes. A Hopf bifurcation is a particular example of this kind of behavioral change. Suppose that we have a system that involves a parameter c,

$$\begin{aligned} x' &= f(x, y, c) \\ y' &= g(x, y, c) \end{aligned} \tag{9}$$

and that has an equilibrium point P at $x = a$, $y = b$ [so that $f(a, b, c) = 0$

and $g(a, b, c) = 0$]. Suppose that the matrix of partial derivatives

☞ This is the Jacobian matrix of system (9). See Chapter 6.

$$J = \begin{bmatrix} \dfrac{\partial f}{\partial x} & \dfrac{\partial f}{\partial y} \\[2ex] \dfrac{\partial g}{\partial x} & \dfrac{\partial g}{\partial y} \end{bmatrix}_{x=a, y=b}$$

has the complex conjugate eigenvalues $\alpha(c) \pm i\,\beta(c)$. The Dutch mathematician Eberhard Hopf showed that *if*:

(a) the functions f and g are twice differentiable,

(b) P is a stable, attracting sink for some value c_0 of the parameter c,

☞ The Hopf conditions.

(c) $\alpha(c_0) = 0$,

(d) $[d\alpha/dc]_{c=c_0} \neq 0$,

(e) $\beta(c_0) \neq 0$,

☞ Since $\alpha'(c_0) \neq 0$, $\alpha(c)$ changes sign as c goes through c_0; this means that P goes from a sink to a source, or the other way around.

then as the parameter c varies through the bifurcation value c_0, the attracting equilibrium point P destabilizes and an attracting limit cycle appears (i.e., an attracting periodic orbit in the xy-phase plane) that grows in amplitude as c changes beyond the value c_0.

It isn't always a simple matter to check the conditions for a Hopf bifurcation (especially condition (b)). It is often easier just to apply the Architect to the system and watch what happens to solution curves and trajectories when a parameter is swept over a range of values. For instance, for system (8) with values $k_2 = 0.08$ and $k_3 = k_4 = 1$ for the rate constants, we can sweep the parameter c and observe the results. In particular, we want to find values of c for which an attracting limit cycle is either spawned by P, or absorbed by P. At and near the special c values we can use the Equilibrium feature of the ODE Architect tool to locate the equilibrium point, calculate the Jacobian matrix, and find its eigenvalues. We expect the eigenvalues to be complex conjugates and the real part to change sign at the bifurcation value of c.

Figure 8.5 shows a sweep of twenty-one trajectories of system (8) with c sweeping down from 1.1 to 0.1 and the values of k_1, k_2, and k_3 as indicated in the figure. See also Problem 3, Exploration 8.4.

✓ (This is the first part of Problem 3 of Exploration 8.4.) Use ODE Architect to duplicate Figure 8.5. Animate (the right icon under Tools on the top menu bar) so that you can see how the trajectories change as c moves downward from 1.1. Then use the Explore feature to determine which values of c spawn or absorb a limit cycle. For what range of values of c does an attracting limit cycle exist?

This behavior of the system (8) carries over to the autocatalator system (7). Notice that the first equation in (7) is $x_1' = -k_1 x_1$, which is easily solved to give $x_1(t) = x_1(0)e^{-k_1 t}$. If k_1 is very small, say $k_1 = 0.002$, the exponential decay of x_1 is very slow, so that if $x_1(0) = 500$, the term $k_1 x_1(t)$,

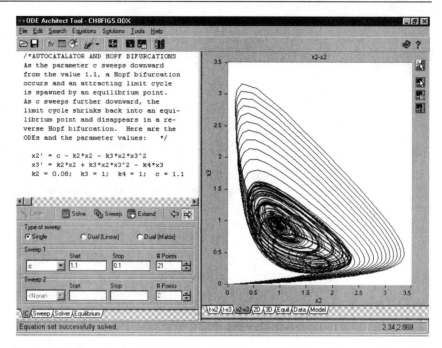

Figure 8.5: Twenty-one trajectories of system (8) for twenty-one values of c; initial data is $x_2(0) = x_3(0) = 0$, time interval is 100 with 1000 points.

though not constant, has values between 1 and 0.01 for a long time interval. The behavior of the autocatalator will be similar to that of system (8).

The section "Bifurcations to a Limit Cycle" in Chapter 7 gives another instance of a Hopf bifurcation. For more on bifurcations, see the references.

References   Gray, P., and Scott, S.K., *Chemical Oscillations and Instabilities* (1990: Oxford Univ. Press)

Hubbard, J.H., and West, B.H., *Differential Equations: A Dynamical Systems Approach, Part II: Higher Dimensional Systems*, (1995: Springer-Verlag)

Scott, S.K., *Chemical Chaos* (1991: Oxford Univ. Press)

Answer questions in the space provided, or on
attached sheets with carefully labeled graphs. A
notepad report using the Architect is OK, too.

Name/Date _____

Course/Section _____

Exploration 8.1. Tracking Pollution in a Lake

1. Suppose that the water flow rates into and out of a lake are $s_{in} = s_{out} = 10^9$ m^3/yr. The (constant) lake volume is $V = 10^{10}$ m^3, and the concentration of pollutant in the water flowing into the lake is $p_{in} = 0.0003$ lb/m^3. Solve the IVP with $L(0) = 0$ (no initial pollution) and describe in words how pollution builds up in the lake. Estimate the steady-state amount of pollution, and estimate the amount of time for the pollution level to increase to half of the asymptotic level.

2. Suppose that the lake in Problem 1 reaches its steady-state level of pollution, and then the source of pollution is removed. Build a new IVP for this situation, and estimate how much time it will take for the lake to clear out 50% of the pollution. How does this time compare to the time you estimated in Problem 1 for the build-up of pollutant?

3.   What would be more effective in controlling pollution in the lake: (**i**) reducing the concentration of pollutant in the inflow stream by 50%, (**ii**) reducing the rate of flow of polluted water into the lake by 50%, or (**iii**) increasing the outflow rate from the lake by 50%?

Answer questions in the space provided, or on
attached sheets with carefully labeled graphs. A
notepad report using the Architect is OK, too.

Name/Date _____

Course/Section _____

Exploration 8.2. What Happens When You Take a Medication?

1. Go to the Library in ODE Architect and check out the file "Cold Pills I: A Model for the Flow of a Single Dose of Medication in the Body" in the folder "Biological Models." This model tracks a unit dose of medication as it moves from the GI tract into the blood and is then cleared from the blood. Read the file and carry out the explorations suggested there. Record your results below.

2. Go to the Library in ODE Architect and check out "Cold Pills II: A Model for the Flow of Medication with Periodic Dosage" in the folder "Biological Models." Carry out the suggested explorations.

3. Suppose you take a decongestant pill every four hours to relieve the symptoms of a cold. Each pill dissolves slowly and completely over the four-hour period between doses, releasing 16 units of decongestant at a constant rate. The decongestant diffuses from the GI tract into the bloodstream at a rate proportional to the amount in the GI tract (rate constant is $a = 0.5/$ hr) and is cleared from the bloodstream at a rate proportional to the amount in the blood (rate constant is $b = 0.1/$ hr). Assume that initially there is no decongestant in the body. Write a report in which you address the following points. Be sure to attach graphs.

(a) Write out ODEs for the amounts $x(t)$ and $y(t)$ in the GI tract and the blood, respectively, at time t.

(b) Find explicit formulas for $x(t)$ and $y(t)$ in terms of $x(0)$ and $y(0)$.

(c) Use ODE Architect to plot $x(t)$ and $y(t)$ for $0 \leq t \leq 100$ hr. What are the equilibrium levels of decongestant in the GI tract and in the blood (assuming that you continue to follow the same dosage regimen)?

(d) Graph $x(t)$ and $y(t)$ as given by the formulas you found in part (b) and overlay these graphs on those produced by ODE Architect. What are the differences?

(e) Imagine that you are an experimental pharmacologist for Get Well Pharmaceuticals. Set lower and upper bounds for decongestant in the bloodstream, bounds that will assure both effectiveness and safety. How long does it take from the time a patient starts taking the medication before the decongestant is effective? How long if you double the initial dosage (the "loading dose")? How about a triple loading dose?

(f) For the old or the chronically ill, the clearance rate constant from the blood may be much lower than the average rate for a random sample of people (because the kidneys don't function as well). Explore this situation and make a recommendation about lowering the dosage.

4. Repeat all of Problem 3 but assume the capsule is rapidly dissolving: it delivers the decongestant at a constant rate to the GI tract in just half an hour, then the dosage is repeated four hours later.

Answer questions in the space provided, or on
attached sheets with carefully labeled graphs. A
notepad report using the Architect is OK, too.

Name/Date ⎯⎯⎯⎯⎯⎯⎯⎯⎯⎯⎯⎯

Course/Section ⎯⎯⎯⎯⎯⎯⎯⎯⎯⎯

Exploration 8.3. Get the Lead Out

1. Check out the ODE Architect Library file "A Model for Lead in the Body" in the "Biological Models" folder and carry out the explorations suggested there. (The notation for the rate constants in the library file differs from the notation used in this chapter.)

2. Use the following rate constants: $k_1 = 0.0039$, $k_2 = 0.0111$, $k_3 = 0.0124$, $k_4 = 0.0162$, $k_5 = 0.000035$, $k_6 = 0.0211$, and put $L = 49.3 \ \mu g/day$ in the lead system (4). These values were derived directly from experiments with volunteer human subjects living in Los Angeles in the early 1970's. Using the data for the lead flow model, describe what happens if the lead inflow rate L is doubled, halved, or multiplied by a constant α. Illustrate your conclusions by using the ODE Architect to graph the lead levels in each of the three body compartments as functions of t. Do the long-term lead levels (i.e., the equilibrium levels) depend on the initial values? On L? Find the equilibrium levels for each of your values of L using ODE Architect. Find the eigenvalues of the Jacobian matrix for each of your values of L. With the names given in Chapter 6 to equilibrium points in mind, would you call the equilibrium lead levels sinks or sources? Nodes, spirals, centers, or saddles?

3. The bones act as a lead storage system, as you can see from the graphs in Submodule 3 of Module 8. What happens if the exit rate constant k_4 from the bones back into the blood is increased from 0.000035 to 0.00035? To 0.0035? Why might an increase in k_4 be harmful? See Problem 2 for the values of L and the rate constants k_i.

4. The medication now in use for acute lead poisoning works by improving the efficiency of the kidneys in clearing lead from the blood (i.e., it increases the value of the rate constant k_6). What if a medication were developed that increased the clearance coefficient k_5 from the tissues? Explore this possibility. See Problem 2 for the values of L and the rate constants k_i.

5. In the 1970's and 80's, special efforts were made to decrease the amount of lead in the environment because of newly enacted laws. Do you think this was a good decision, or do you think it would have been better to direct the efforts toward the development of a better antilead medication for cases of lead poisoning? Why? What factors are involved in making such a decision?

Answer questions in the space provided, or on attached sheets with carefully labeled graphs. A notepad report using the Architect is OK, too.

Name/Date _____

Course/Section _____

Exploration 8.4. Chemical Reactions: the Autocatalator

1. Check out "The Autocatalator Reaction" in the "Chemical Models" folder in the ODE Architect Library and graph the concentrations suggested. Describe how the concentrations of the various chemical species change in time.

2. Here are schematics for chemical reactions. Draw a compartment diagram for each reaction. Then write out the corresponding sets of ODEs for the individual chemical concentrations. [Use lower case letters for the concentrations (e.g., $x(t)$ for the concentration of chemical X at time t).]

   (a) $X + Y \xrightarrow{k} Z$

   (b) $X + Y \xrightarrow{k_1} Z \xrightarrow{k_2} W$

   (c) $X + 2Y \xrightarrow{k} Z$

   (d) $X + 2Y \xrightarrow{k} 3Y + Z$

3. Explore the behavior of $x_2(t)$ and $x_3(t)$ as governed by system (8). Start with $c = 1.1$, $k_2 = 0.08$, $k_3 = k_4 = 1$, and $x(0) = y(0) = 0$. Then sweep c from 1.1 down to 0.1 and describe what happens to orbits in the x_2x_3-plane. Find the range of values of c between 1.1 and 0.1 for which attracting limit cycles are visible. These are Hopf cycles. Fix c at a value that you think is interesting and sweep one the parameters k_2, k_3, or k_4. Describe what you observe. [*Suggestion:* Take a look at Figure 8.5, and use the Animate feature of ODE Architect to scroll through the sweep trajectories. Then use the Explore option to get a data table with information about any of the trajectories you have selected.]

4. Look at the autocatalator system (7) with $x_1(0) = 500$, $x_2(0) = x_3(0) = x_4(0) = 0$ and $k_1 = 0.002$, $k_2 = 0.08$, $k_3 = k_4 = 1$. Graph $x_2(t)$ and $x_3(t)$ over various time ranges and estimate the times when sustained oscillations begin and when they finally stop. What are the time intervals between successive maxima of the oscillations in x_2? Plot a 3D tx_2x_3-graph over various time intervals ranging from $t = 100$ to $t = 1000$. Describe what you see. [*Suggestion:* Look at the chapter cover figure and Figure 8.4.]

☞ Use the Explore feature to estimate the starting and stopping times of the oscillator.

     Now sweep the values of $x_1(0)$ downward from 500. What is the minimal value that generates sustained oscillations? Then fix $x_1(0)$ at 500 and try to turn the oscillations off by changing one or more of the rate constants k_1, k_2, k_3, k_4—this corresponds to heating or chilling the reactor. Describe your results.

9 Population Models

CHAPTER

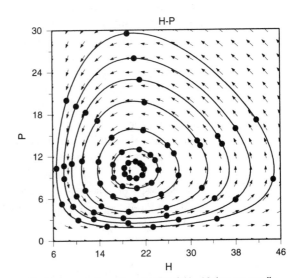

Predator-prey cycles with a direction field. Markers are equally spaced in time.

Overview   Population biology is the study of how communities of organisms change. The structure of a population can be quite intricate, such as species interactions in a tropical rain forest. Other communities may involve only a few species and are simpler to describe. There are many aspects of population biology, including ecology, demography, population genetics, and epidemiology. In each of these areas, mathematics plays an important role in modeling how populations change in time and how the interaction between the environment and the community affects that change. We'll explore mathematical models in ecology and epidemiology.

Key words   Logistic model; growth rate; carrying capacity; equilibrium; steady state; competition; coexistence; exclusion; predator-prey; epidemiology

See also   Chapter 1 for more on the logistic equation, Chapter 7 for predator-prey models, and Chapter 8 for chemical mass action.

◆ Modeling Population Growth

The increasing awareness of environmental issues is an important development in modern society. This awareness ranges from concern about conserving important natural resources to concern about habitat destruction and the endangerment of species. Human population pressures are ever-increasing, and this growth has led to intense exploitation of the environment. To reduce the negative effects of this exploitation, scientists are seeking to understand the ecology and biology of natural populations. This understanding can be used to design management strategies and environmental policies.

The simplest ecological models describe the growth of a single species living in an *environment* or *habitat*. Characteristics of the habitat—moisture, temperature, availability of food—will affect how well the species survives and reproduces. Intrinsic biological characteristics of the species, such as the basic reproductive rate, also affect the growth of the species. However, a mathematical model that incorporates *all* possible effects on the growth of the population would be complicated and difficult to interpret.

✓   "Check" your understanding by answering this question: What are some other characteristics of a species and its environment that can affect the productivity of the species?

The most common procedure for modeling population growth is first to build elementary models with only a few biological or environmental features. Once the simple models are understood, more complicated models can be developed. In the next section we'll start with the *logistic model* for the growth of a single species—a model that is both simple and fundamental.

◆ The Logistic Model

The ecological situation that we want to model is that of a single species growing in an environment with limited resources. Examples of this situation abound in nature: the fish population of a mountain lake (the limited resource is food), a population of ferns on a forest floor (the limited resource is light), or the lichen population on a field of arctic rocks (the limited resource is space). We won't attempt to describe the biology or ecology of our population in detail: we want to keep the mathematical model simple. Instead, we'll summarize a number of such effects using two *parameters*. The first parameter is called the *intrinsic growth rate* of the population. It is often symbolized using the letter r, and it represents the average number of offspring, per unit time, that each individual contributes to the growth of the population. The second parameter is called the *carrying capacity of the environment*. Symbolized by K, the carrying capacity is the largest number of individuals that the environment can support in a steady state. If there are more individuals in the population than the carrying capacity, the population declines because there are too few resources to support them. When there are fewer individuals than K, the environment has not been overexploited and the population grows.

Our mathematical model must capture these essential characteristics of the population's growth pattern. To begin, we define a variable that represents the size of the population as a function of time; call this size $N(t)$ at time t. Next, we specify how the size $N(t)$ changes in time. Creating a specific rule for the rate of change of population size is the first step in *building a mathematical model*. In general, a model for changing population size has the form

$$\frac{dN}{dt} = f(t, N), \quad N(0) = N_0$$

for some function f and initial population N_0. Once the details of f are given, based on the biological and ecological assumptions, we have a concrete mathematical model for the growth of the population.

To complete our description of the logistic model, we need to find a reasonable function f that captures the essential properties described above. We are looking for a simple function that gives rise to

- population growth when the population is below the carrying capacity (that is, $N'(t) > 0$ if $N(t) < K$);

- population decline if the population exceeds the carrying capacity (that is, $N'(t) < 0$ if $N(t) > K$).

One such function is $f(t, N) = rN(1 - N/K)$. The *logistic model* is the IVP

$$\frac{dN}{dt} = rN\left(1 - \frac{N}{K}\right), \quad N(0) = N_0$$

where r, K and N_0 are positive constants. Figure 9.1 shows some typical solution curves.

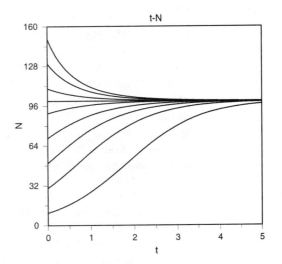

Figure 9.1: Some solution curves for the logistic equation.

Let's observe a few important features of our model. First, the algebraic sign of $N'(t)$ follows the desired relationship to the population size N. Second, if either $N = 0$ or $N = K$, there is no change in the population size: $N'(t) = 0$. Thus, $N = 0$ and $N = K$ are *equilibria* or *steady states*. The first steady state corresponds to the extinction of the species in its environment, and the second corresponds to a population in perfect balance, living at the carrying capacity of the habitat.

Notice the effect of the parameters r and K. As the carrying capacity K increases, the environment supports more individuals at equilibrium. As the growth rate r increases, the population attains its steady state in a faster time. An important part of understanding a mathematical model is to discover how changing the parameters affects the behavior of the system that is being modeled. This knowledge can lead to predictions about the system, and to a much deeper understanding of population processes. In Exploration 9.1 you will study the logistic model and variations of it. See also Chapter 1.

✓   Do you think the intrinsic annual growth rate r of the earth's human population is closer to 0.01, 0.03, or 0.05? It's anyone's guess as to the carrying capacity. What is your estimate, given that the current population is about 6 billion?

◆ Two-Species Population Models

The logistic model applies to a single species. Most habitats support a variety of species; interactions can be both *intraspecific* (between individuals of the same species) or *interspecific* (between individuals of different species). These interactions can take many forms. For example, competition between individuals of the same species for food resources, nesting sites, mates, and so on, are intraspecific interactions that lead to regulated population growth. Important interspecific interactions include predation, competition for food or other resources, and symbiotic relationships that are mutually beneficial. Such interactions can be very complex and can involve a large number of species. Again, the first step in modeling complicated ecologies is to build and analyze simple models. We'll present two such models here (involving only two species) and consider others in the last three explorations.

✓   Can you think of a mutually beneficial interaction between humans and another species?

◆ Predator and Prey

As we noted, an important interaction between species is that of predator and prey. Such interactions are very common: animals must eat to thrive, and for every eater there is the eaten! Spiders prey on flies, cows prey on grass, mosquitoes prey on humans, and humans prey on shiitake mushrooms, truffles, salmon, redwood trees, and just about everything else. We'll now build a simple model to describe such interactions.

Consider two species, the prey species (H, because they're "harvested" or "hunted") and the predator species (P), but for the moment, imagine that they don't interact. In the absence of the predator, we assume that the prey grows according to the logistic law, with carrying capacity K and intrinsic growth rate $a > 0$. The model for the prey under these conditions is

$$H' = aH(1 - H/K)$$

Now suppose that in the absence of its food source (the prey), the predator dies out; the model for the predator is

$$P' = -bP$$

where $b > 0$. If this situation persists, the prey will grow to fill the habitat and the predator will become extinct.

Now suppose that the predator does feed upon the prey, and that each predator consumes, on the average, a fraction c of the prey population, per unit time. The growth rate of the prey will then be decreased (since they're being eaten) by the amount cHP. The predators benefit from having consumed the prey, so their growth rate will increase. But because a given predator may have to consume a lot of prey to survive, not all prey produce new predators in a one-for-one way. Therefore the increase in the growth rate of the predators in this case is dHP, where d is a constant which may be different than c. Putting this all together, we obtain our model for the predator-prey system:

☞ When two species interact at a rate proportional to the product of the two populations, it's called *population mass action*.

$$\frac{dH}{dt} = aH\left(1 - \frac{H}{K}\right) - cHP$$
$$\frac{dP}{dt} = -bP + dHP \tag{1}$$

Analyzing this model gives insight into a number of important ecological issues, such as the nature of coexistence of predator and prey, and the understanding of population cycles. Figure 9.2 on the next page shows a phase plot for the predator-prey system described by ODE (1). Exploration 9.3 examines this predator-prey model.

✓ What is the long-term future of the prey species in Figure 9.2? The predator species?

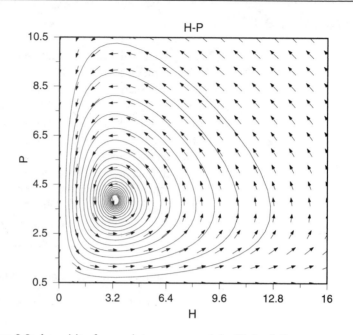

Figure 9.2: An orbit of a predator-prey model with logistic prey growth.

◆ Species Competition

Another common interaction between species is *competition*. Species can compete for space, food, light, or for other resources. In the absence of its competitor, each species grows logistically to its carrying capacity. However, the presence of the competitor changes the situation, and the growth rate of each species is diminished by the presence of the other. Let N_1 and N_2 represent the numbers of the two species. We model the competition between these species with the following equations:

$$\frac{dN_1}{dt} = r_1 N_1 \left(1 - \frac{N_1}{K_1} - \alpha_{12} N_2 \right)$$

$$\frac{dN_2}{dt} = r_2 N_2 \left(1 - \frac{N_2}{K_2} - \alpha_{21} N_1 \right) \tag{2}$$

The parameter α_{12} measures the effect of Species 2 on Species 1, and α_{21} measures the effect of Species 1 on Species 2. If $\alpha_{12} > \alpha_{21}$, then Species 2 dominates Species 1, because Species 2 reduces the growth rate of Species 1 more *per capita* than the reverse. The analysis of this model gives insight into how species maintain their diversity in the ecology (*coexistence*) or how such diversity might be lost (*competitive exclusion*). A phase plot for the competitive system is shown in Figure 9.3. Exploration 9.4 examines a related model for so-called mutualistic interactions.

☞ Could the competition be so fierce that both species become extinct?

✓ Does Figure 9.3 show coexistence or competitive exclusion?

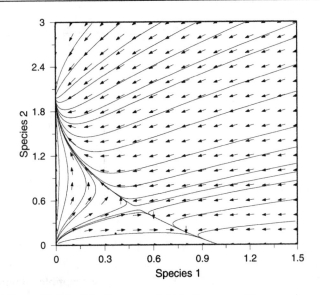

Figure 9.3: Orbits for a model of two competing species.

◆ Mathematical Epidemiology

An important use of mathematical models is to describe how infectious diseases spread through populations. This field is called epidemiology. Quantitative models can predict the time course of a disease or the effectiveness of control strategies, such as immunization. Again, the development proceeds from the simplest model to more complex ones.

The most elementary model for an epidemic is the so-called SIR model (presented in Module 9): Consider a population of individuals divided into three groups—those susceptible (S) to a certain disease, those infected (I) with the disease , and those who have recovered (R) and are immune to reinfection, or who otherwise leave the population. The SIR model describes how the proportions of these groups change in time.

The susceptible population changes size as individuals become infected. Let's think of this process as "converting" susceptibles to infecteds. If we assume that each infected individual can infect a proportion a of the susceptible population per unit time, we obtain the rate equation

☞ Another instance of population mass action.

$$\frac{dS}{dt} = -aSI \qquad (3)$$

The infected population is increased by conversion of susceptibles and is decreased when infected individuals recover. If b represents the proportion of infecteds that recover per unit time, then the rate of change of the infected population satisfies

$$\frac{dI}{dt} = aSI - bI \qquad (4)$$

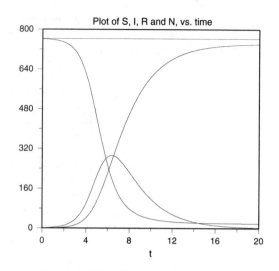

Figure 9.4: A plot of susceptibles (falling curve), infecteds (the "bump"), re-coloreds (rising curve), and their sum (top line).

Lastly, as infecteds recover they augment the recovered population, so that

$$\frac{dR}{dt} = bI \tag{5}$$

The ODEs (3)–(5) together with the initial values $S(0)$, $I(0)$, and $R(0)$ define the SIR model. A component plot of S, I, R, and $N = S + I + R$ appears in Figure 9.4.

✓ Can you explain why $N(t)$ stays constant as time changes?

We can learn many important things from this model about the spread of diseases. For example, analysis of the model can reveal how the rate of spread of the disease through the population is related to the infectiousness of the disease. Our common experience suggests that not all diseases become epidemic: sometimes a few people are afflicted and the disease dies out. Analysis of the SIR model can also give insight into the conditions under which a disease will become epidemic. This is an important phenomenon! These and other matters will be examined in Exploration 9.5.

✓ What factors can you think of that might influence the spread of a disease in a human population?

References   Bailey, N.T.J., *The Mathematical Theory of Infectious Diseases and its Applications*, 2nd ed., (1975: Hafner Press)

Edelstein-Keshet, L., *Mathematical Models in Biology*, (1988: McGraw-Hill)

Answer questions in the space provided, or on
attached sheets with carefully labeled graphs. A
notepad report using the Architect is OK, too.

Name/Date _____

Course/Section _____

Exploration 9.1. The Logistic Model

In this exploration you will consider a population that grows according to the logistic law: $N' = rN(1 - N/K)$, where r is the intrinsic growth rate and K is the carrying capacity of the environment.

☞ Using the ODE
Architect Library.

1. Open the ODE Architect Library. In the folder "Population Models," open the file "Logistic Model of Population Growth." The logistic equation will be automatically entered into the Architect. The graphs show several solution curves. Set the initial condition for the population size to $N_0 = 25$ and set $K = 100$. Plot eight solutions by sweeping the growth rate constant from $r = -0.5$ to $r = 2$; print your graph. Describe the effect of r on the solutions of the logistic equation. Your description should address the following questions: How does the growth rate constant affect the long-term behavior of the population? How does the rate constant affect the dynamics of the system?

2. Set the IC to $N_0 = 25$ and $r = 1.2$. Plot eight solution curves by sweeping the carrying capacity K from 70 to 150; print your graph. Describe the effect of the parameter K on the solutions of the logistic equation. Your description should address the following questions: How does the carrying capacity affect the long-term behavior of the population? How does the carrying capacity affect the dynamics of the system?

3. Study the graphs that you produced for Problems 1 and 2. Notice that sometimes the rate of change of population size is increasing (i.e., $N'(t)$ is increasing and the graph of $N(t)$ is concave up) and sometimes it is decreasing ($N'(t)$ is decreasing and the graph of $N(t)$ is concave down). By analyzing your graphs, try to predict a relationship between r, K, and N that distinguishes between these two situations. Use ODE Architect to test your prediction by graphing more solution curves. Lastly, try to confirm your prediction by exact analysis of N'' using the logistic ODE.

Answer questions in the space provided, or on
attached sheets with carefully labeled graphs. A
notepad report using the Architect is OK, too.

Name/Date _____

Course/Section _____

Exploration 9.2. Harvesting a Natural Resource

Human societies use resources from their environments. We harvest animals
and plants for food, construction, fuel, and many other uses. The harvesting
of a biological resource must be done carefully, because overexploitation of
the population can cause severe harm, or even extinction, to the resource. As
a society we have become much more sensitive about the need to balance
the benefits of resource consumption against the impact of that consumption
on the exploited populations and their environment.

Resource management is an important tool for minimizing the negative
effects of harvesting. Mathematical models are tools for understanding the
impact of harvesting on a population, so that we can then design management policies, such as quotas on the annual harvest.

In this exploration, you will analyze a simple model for harvesting a single species. To be specific, suppose that the habitat is a forest and the resource is a species of pine tree. The number of trees grows logistically with
an intrinsic growth rate r, and the forest will support at most K trees (measured in millions of board feet). You are a consulting ecologist, asked to
model the effect of a lumber company's harvesting strategy on the pine forest. The company harvests the trees *proportionally*: in a unit of time (a year,
for example), the company removes a fixed fraction h of the trees. Harvesting reduces the net rate of growth of the forest; this leads you to propose the
following model for the effect of harvesting:

$$N' = rN\left(1 - \frac{N}{K}\right) - hN, \quad N(0) = N_0 \tag{6}$$

The last term, $-hN$, is the *harvesting term*. Notice that when $h = 0$ (i.e., no
trees are harvested), the model reduces to the logistic equation.

1. Open ODE Architect and enter the ODE for the harvesting model given by
   equation (6). Set the growth rate to $r = 0.1$ year^{-1}, the carrying capacity to
   $K = 1000$ million board feet, and the population size IC to $N_0 = 100$ at $t = 0$.
   Describe the growth of the forest when there is no harvesting ($h = 0$). You'll
   have to choose a good time interval to best display your results.

2. Keep r and K fixed and plot solution curves for various (positive) values of the harvesting rate h. You can do this exploration most efficiently by sweeping the parameter h. After you have studied a variety of harvest rates, explain how harvesting affects the pine population. Your explanation should address the following questions: How does the growth of the pine population with harvesting compare to its growth without harvesting? What is the long-term effect of harvesting? How are the time dynamics of the forest growth affected by harvesting?

3. The annual yield Y of the harvest is the amount of lumber removed per year. This is just $Y = hN$ when there are N units of lumber in the forest. The yield will vary through time as the amount of lumber (trees) in the forest varies in time. If the harvest rate is too high, the long-term yield will tend to zero ($Y \to 0$) and the forest will become overexploited. If the harvest rate is very low, the yield will also be very low. As the consultant to the company, you are asked: What should the harvest rate be to give the largest *sustainable yield* of lumber? That is to say, what optimal harvest rate will maximize $\lim_{t \to \infty} Y(t)$? Attack the problem graphically using ODE Architect to plot graphs of the yield function for various values of h. Assume that $r = 0.1$, $K = 1000$, and $N_0 = 100$. If you can, provide an analytic solution to the question, and check your results using the Architect. Suppose that the company follows your recommendation and harvests pine at the optimal rate. When the size of the forest reaches equilibrium, how much lumber (trees) will there be, and how does this amount compare to the size of the forest without harvesting?

Answer questions in the space provided, or on
attached sheets with carefully labeled graphs. A
notepad report using the Architect is OK, too.

Name/Date _____

Course/Section _____

Exploration 9.3. Predator and Prey

Predator-prey interactions are very common in natural populations. These interactions can be modeled by a system of nonlinear equations:

$$H' = aH\left(1 - \frac{H}{K}\right) - cHP, \quad P' = -bP + dHP$$

where H and P are the prey and predator population sizes, respectively.

1. Give a biological interpretation of the parameters a, b, c, d, K of the predator-prey model.

2. Open ODE Architect and enter the modeling equations for the predator-prey system above. Assign the following values to the parameters: $a = 1.0$, $b = 0.25$, $c = 1.0$, $d = 0.15$, $K = 100$. After you have entered the equations and parameters, set the solve interval to 60, and the number of points plotted to 500. Solve the system forward in time using the initial conditions $H(0) = 1$, $P(0) = 1$. Plot graphs of the orbits in the HP-phase plane, and plot the individual component graphs for predator and prey. Experiment with other initial conditions. Describe the nature of the solutions and locate all equilibrium solutions.

3.   Fix $a = 1.0$, $b = 0.25$, $c = 1.0$, and $d = 0.15$ as in Problem 2. Plot several solutions from the fixed initial conditions $H(0) = 1$, $P(0) = 1$, for varying values of K. For example, let K range over several values between 100 to 10,000. How does changing the carrying capacity of the prey affect the behavior of the system? Make a conjecture about the limiting behavior of the system as $K \to \infty$.

4.   Test the conjecture that you made in Problem 3 in two steps:

(a) Take the limit as $K \to \infty$ in the predator-prey equations and obtain a new system of equations that describes a predator-prey system where there is no resource limitation for the prey.

(b) Explore this system using ODE Architect; this new system is often called the *Lotka–Volterra* model. Plot several orbits using markers that are equally spaced in time. Do the cycles have a common period? How do the time markers help you answer that question? Compare your graphs with the chapter cover figure. Also plot graphs of H against t and P against t for various values of $H(0)$ and $P(0)$. What do these graphs tell you about the periods?

☞ Finally, you get a chance to figure out what is going on in the chapter cover figure.

How does the behavior of the Lotka–Volterra model differ from the model you explored in Problems 1–3?

Answer questions in the space provided, or on attached sheets with carefully labeled graphs. A notepad report using the Architect is OK, too.

Name/Date _____

Course/Section _____

Exploration 9.4. Mutualism: Symbiotic Species Interactions

For both predator-prey and species competition, the growth rate of at least one of the species is reduced by the interaction. Though eating the prey helps the predator, it certainly harms the prey; for competitors the reduction in growth rate is reciprocal. Not all species interactions must be negative: there are many examples where the species cooperate or otherwise *mutually enhance* their respective growth rates. A famous example is the yucca plant–yucca moth system: the yucca plant can be pollinated only by the yucca moth, and the yucca moth is adapted to eat nectar only from the yucca plant. Each species benefits the other, and their interaction is positive for both. Such interactions are called *mutualistic* or *symbiotic* by ecologists. In this exploration we will present and analyze a simple model for mutualism.

Our model will be very similar to the competition model studied in Module 9. To obtain a model for mutualism, we just change the signs of the interaction terms so that they are always positive: each species enhances the growth rate of the other. We then obtain the following equations:

$$\frac{dN_1}{dt} = N_1 \left(r_1 - e_1 N_1 + \alpha_{12} N_2 \right), \quad \frac{dN_2}{dt} = N_2 \left(r_2 - e_2 N_2 + \alpha_{21} N_1 \right) \qquad (7)$$

The parameters r_1, r_2, α_{12}, and α_{21} retain their meanings from ODE (2) in the competition model. However, the interaction terms $\alpha_{12} N_1 N_2$ and $\alpha_{21} N_1 N_2$ have *positive* sign and thus enhance the respective growth rates.

Notice that in the absence of interaction, the carrying capacities of the two species are $K_1 = r_1/e_1$ and $K_2 = r_2/e_2$ in this version of the model.

1. Open the ODE Architect Library, go to the "Population Models" folder, and open the file "Mutualism: Symbiotic Interactions." This file loads the equations that model a mutualistic interaction. Fix $r_1 = 1$, $r_2 = 0.5$, $e_1 = 1$, $e_2 = 0.75$. Vary the values of each of the interaction coefficients from 0 to 2. For each combination of values for α_{12} and α_{21} that you try, draw a phase portrait of the system (7) in the first quadrant. Describe every possible kind of behavior of the system; try enough combinations of the parameters to feel confident that you have covered all the possibilities. Answer the following questions: Is species coexistence possible? Can competitive exclusion occur? Will the populations of both species remain bounded as time increases?

☞ You may want to use the Dual (Matrix) sweep feature here.

2. Using pencil and paper, deduce the conditions under which a two-species equilibrium will be present. Check your conditions using the Architect to solve the model. When a two-species equilibrium is present, does it necessarily have to be stable? Compare two-species equilibria to single-species equilibria (the carrying capacities): does mutualism increase or decrease the abundance of the species at equilibrium?

3. Do you think that a mutualistic interaction is always beneficial to an ecosystem? Under what conditions might it be deleterious? Compare the behavior of mutually interacting species to that of competing species. How are the two behaviors similar? How are they different?

Answer questions in the space provided, or on
attached sheets with carefully labeled graphs. A
notepad report using the Architect is OK, too.

Name/Date _____

Course/Section _____

Exploration 9.5. Analyzing the SIR model for an Epidemic

We will now explore the SIR model for the spread of an epidemic. Recall the
ODEs for this model: $S' = -aSI$, $I' = aSI - bI$, $R' = bI$. The parameter $a > 0$ is
the infection rate constant and $b > 0$ is the removal (recovery) rate constant
of infecteds. Notice that $S' + I' + R' = 0$, i.e., the total number of individuals
N is constant and equals $S(0) + I(0) + R(0)$. The ODE Architect Library has an
equation file for the SIR model in the "Population Models" folder. In this file
you will find values of a, b, and N that correspond to an actual epidemic.

1. Set the IC to $I(0) = 20$ and $R(0) = 0$. Set the solve interval to 24 time
   units, and make ten plots by sweeping the initial number of susceptibles from
   $S(0) = 100$ to $S(0) = 500$. Now examine the graph panel for I vs. t. Which
   of the curves corresponds to $S(0) = 100$ and which to $S(0) = 500$? By definition, an epidemic occurs if $I(t)$ increases from its initial value $I(0)$. For
   which of the curves that you plotted did an epidemic occur?

2. The behavior that you studied in Problem 1 is called a *threshold effect*. If the
   initial number of susceptible individuals is below a threshold value, there will
   be no epidemic. If the number of susceptibles exceeds this value, there will
   be an epidemic. Use ODE Architect to empirically determine the threshold
   value for $S(0)$; use the values of a and b in the Library file. Now analyze the
   equation for dI/dt and determine a sufficient condition for $I(t)$ to initially
   increase. Interpret your answer as a threshold effect. Use the values of the
   infection and removal rates that appear in the Library file to compare your
   analytic calculation of the threshold with that obtained from your empirical
   study.

3. Clear your previous results from the Architect but keep the same values for a and b. Set the initial conditions for I, R, and S to $10, 0$, and 200, respectively. Solve the equations for a time interval of 24 units. Notice from the plot of $I(t)$ that the number of infecteds steadily diminishes from 10 to nearly zero. Also notice that over this same period of time, the number of susceptibles declines by almost 50, and the number of recovered individuals increases from zero to nearly 50. Explain this seemingly contradictory observation.

4. A disease is said to be *endemic* in a population if at equilibrium the number of infecteds is positive. Is it possible in the SIR model for the disease to be endemic at equilibrium? In other words, can $\lim_{t \to \infty} I(t) > 0$? Explain your answer.

10 The Pendulum and Its Friends

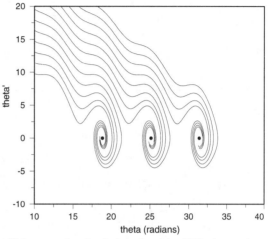

High-energy trajectories of a damped pendulum ODE swing over the top and then settle into decaying oscillations about rest points.

Overview   The whole range of fixed-length pendulum models—linear, nonlinear, damped, and forced—are presented in this chapter, and their behaviors are compared using insights provided by integrals. After discussing fixed-length pendulum ODEs, the effects of damping, and separatrices, we turn to a variable-length model. A child pumping a swing alters the length of its associated pendulum as the swing moves. We present a nontraditional autonomous model and show that phase-plane analysis leads to a successful description of the effects of the pumping action. Finally, the problem of finding geodesics (the paths of minimum length between points) on a torus leads to an ODE with a striking resemblance to the pendulum ODE.

Key words   Linear pendulum; nonlinear pendulum; damping; energy; pumping (a swing); conservation laws; torus; geodesic; limit cycle; bifurcation

See also   Chapter 4 for a spring-mass system which has the same ODE as the linear pendulum; Chapter 11 for a study of damping effects in the Robot and Egg submodule, and a lengthening pendulum in Exploration 11.4; and Chapter 12 for elaboration on the forced, damped pendulum resulting in chaos (and control).

◆ Modeling Pendulum Motion

☞ The volumes by Halliday and Resnick (refs.) are good general references for physical models (including the pendulum).

Let's find the ODE that models the motion of a pendulum. For a pendulum bob of mass m at the end of a rod of negligible weight and fixed length L at an angle θ to the vertical, Newton's second law gives

$$\text{mass} \cdot \text{acceleration} = \text{sum of forces acting on the bob}$$

The bob moves along an arc of a circle of radius L. The tangential component of the bob's velocity and acceleration at time t are given by $L\theta'(t)$ and

☞ Since the tensile force in the rod and the radial component of the gravitational force are equal and opposite, the radial acceleration is zero and the pendulum moves along a circular arc.

$L\theta''(t)$, respectively. The tangential component, $-mg\sin\theta$, of the gravitational force acts to restore the pendulum to its downward equilibrium. The viscous damping force, $-bL\theta'$, is proportional to the velocity and acts in a direction tangential to the motion, but oppositely directed. Other forces such as repeated pushes on the bob may also have components $F(t)$ in the tangential direction.

Equating the product of the mass and the tangential acceleration to the sum of the tangential forces, we obtain the pendulum ODE

$$mL\theta'' = -mg\sin\theta - bL\theta' - F(t) \tag{1}$$

The equivalent pendulum system is

$$\begin{aligned}
\theta' &= y \\
y' &= -\frac{g}{L}\sin\theta - \frac{b}{m}y + \frac{1}{mL}F(t)
\end{aligned} \tag{2}$$

The angle θ is positive if measured counterclockwise from the downward vertical, and is negative otherwise; θ is measured in radians (1 radian is $360/2\pi$ or about $57°$). We allow θ to increase or decrease without bound because we want to keep track of the number of times that the pendulum swings over the pivot, and in which direction. For example, if $\theta = -5$ radians then the pendulum was swung clockwise (the minus sign) once over the top from $\theta = 0$ because the angle -5 is between $-\pi$ (at the top clockwise from 0) and -3π (reaching the top a second time going clockwise).

We will work with the *undriven* pendulum ODE ($F = 0$) in this chapter. Since $\sin\theta \approx \theta$ if $|\theta|$ is small, we will on occasion replace $\sin\theta$ by θ to obtain a linear ODE. We treat both undamped ($b = 0$) and damped ($b > 0$) pendulum ODEs:

☞ The first two ODEs in this list have the form of the mass-spring ODEs of Chapter 4.

$$\theta'' + \frac{g}{L}\theta = 0 \quad \text{(undamped, linear)} \tag{3a}$$

$$\theta'' + \frac{b}{m}\theta' + \frac{g}{L}\theta = 0 \quad \text{(damped, linear)} \tag{3b}$$

$$\theta'' + \frac{g}{L}\sin\theta = 0 \quad \text{(undamped, nonlinear)} \tag{3c}$$

$$\theta'' + \frac{b}{m}\theta' + \frac{g}{L}\sin\theta = 0 \quad \text{(damped, nonlinear)} \tag{3d}$$

$$\theta'' + \frac{b}{m}\theta' + \frac{g}{L}\sin\theta = \frac{1}{mL}F(t) \quad \text{(damped, nonlinear, forced)} \tag{3e}$$

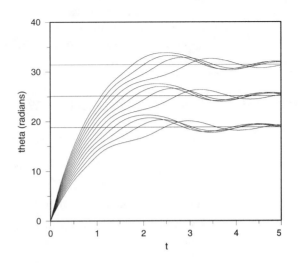

Figure 10.1: Solution curves of a damped pendulum system. What is the meaning of the horizontal solution curves?

Figure 10.1 and the chapter cover figure, respectively, show some solution curves and trajectories of the damped, nonlinear pendulum ODE, $\theta'' + \theta' + 10 \sin \theta = 0$. Although the two linear ODEs are only good models of actual pendulum motions when $|\theta|$ is small, these ODEs have the advantages that their solutions have explicit formulas (see Chapter 4). The nonlinear ODEs model pendulum motions for all values of θ, but there are no explicit solution formulas.

Now fire up your computer, go to Screen 1.2 of Module 10, and visually explore the behavior of solution curves and trajectories of linear, nonlinear, damped, and undamped pendulum ODEs. Pay particular attention to the behavior of the animated pendulum at the upper left, and relate its motions to the trajectories and to the solution curves, and to what you think a real pendulum would do. Explore all the options in order to understand the differences.

✓   "Check" your understanding by matching solution curves of Figure 10.1 with the corresponding trajectories in the chapter cover figure. Describe the long-term behavior of the pendulum represented by each curve.

☞ This is also what Problem 1 of Exploration 10.1 is about.

✓   Go to Screen 1.2 of Module 10 and explore what happens to solutions of the undamped, linearized ODE, $\theta'' + \theta = 0$, if θ_0 is 0 and θ_0' is large. The motion of the animated pendulum is crazy, even though it accurately portrays the behavior of the solutions $\theta(t) = \theta_0' \sin t$. Explain what is going on. Is the linearized ODE a good model here? Repeat with the undamped, nonlinear ODE, $\theta'' + \sin \theta = 0$, and the same initial data as above. Is this a better model?

There is another way to look at pendulum motion, an approach based on integrals of motion. This approach goes beyond pendulum motion and applies to any physical system which can be modeled by a second-order ODE of a particular type.

◆ Conservative Systems: Integrals of Motion

In this section we will study solutions of the differential equation

☞ V is often called a potential function.

$$q'' = -\frac{dV}{dq} \tag{4}$$

for a generic variable q where $V(q)$ is a given function.

Example 1: The undamped, nonlinear pendulum ODE is the special case where $q = \theta$:

$$\theta'' = -\frac{g}{L}\sin\theta, \qquad V(\theta) = -\frac{g}{L}\cos\theta$$

Example 2: You will see later in this chapter that geodesics on a surface of revolution lead to the differential equation

$$u'' = M^2\frac{f'}{f^3}, \qquad V(u) = M^2\frac{1}{2f^2}$$

where the generic variable q is u in this case, M is a constant, and f is a function of u.

ODE (4) is autonomous and equivalent to the system

$$\begin{aligned} q' &= y \\ y' &= -\frac{dV}{dq} \end{aligned} \tag{5}$$

A solution to system (5) is a pair of functions, $q = q(t)$, $y = y(t)$. One way to analyze the behavior of these solutions is by a conservation law. A function $K(q, y)$ that remains constant on each solution [i.e., $K(q(t), y(t))$ is a constant for all t], but varies from one solution to another, is said to be a *conserved quantity*, or an *integral of motion* and the system is said to be *conservative*. For system (5) one conserved quantity is

$$K(q, y) = \frac{1}{2}y^2 + V(q) \tag{6}$$

Here's how to prove that $K(q(t), y(t))$ stays constant on a solution—use the chain rule and system (5) to show that dK/dt is zero:

$$\frac{dK}{dt} = y\frac{dy}{dt} + \frac{dV}{dq}\frac{dq}{dt} = y\left(-\frac{dV}{dq}\right) + \frac{dV}{dq}y = 0$$

Incidentally, if K is any conserved quantity, so also is $\alpha K + \beta$ where α and β are constants and $\alpha \neq 0$.

Example 3: Here's an example where we use $\alpha K + \beta$, rather than K, as the integral to show how integrals sometimes correspond to physical quantities. Look back at the function $V(\theta) = -(g/L)\cos\theta$ for the undamped, nonlinear pendulum of Example 1. Using formula (6), $E(\theta, y)$ is an integral, where

$$E(\theta, y) = mL^2 K(\theta, y) + mgL$$

$$= mL^2 \left(\frac{1}{2}y^2 - \frac{g}{L}\cos\theta\right) + mgL$$

$$= \frac{1}{2}m(Ly)^2 + mgL(1 - \cos\theta)$$

$$= \text{kinetic energy} + \text{potential energy}$$

This integral is called the *total mechanical energy* of the pendulum. The constant mgL is inserted so that the potential energy is zero when the pendulum bob is at its lowest point.

✓   Find the conserved quantity E for the undamped, linear pendulum ODE $\theta'' + \theta = 0$. Draw level curves $E(\theta, y) = E_0$, where $y = \theta'$, in the θy-plane, and identify the curves (e.g., ellipses, parabolas, hyperbolas).

☞ So we can draw trajectories of system (5) by drawing level sets of an integral.

Drawing the level curves of a conserved quantity K in the qy-plane for system (5) gives phase plane trajectories of the system and so serves to describe the motions. This may be much easier than finding solution formulas, but even so, we can take some steps toward obtaining formulas. To see this, we have from equation (6) that if K has the value K_0 on a trajectory of system (5), then

$$\frac{1}{2}y^2 + V(q) = K_0, \quad \text{i.e.,} \quad y = q' = \pm\sqrt{2K_0 - 2V(q)}$$

This is a separable first-order differential equation (as discussed in Chapter 2) that can be solved by separating the variables and integrating:

$$\int \frac{dq}{\sqrt{K_0 - V(q)}} = \sqrt{2}t + C$$

It is hard to obtain explicit solution formulas because the integral cannot usually be expressed in terms of elementary functions.

◆ The Effect of Damping

Mechanical systems are usually damped by friction, and it is important to understand the effect of friction on the motions. Friction is not well described by the fundamental laws of physics, and any formula we write for it will be more or less *ad-hoc*. The system will now be modeled by a differential equation of the form

$$q'' + f(q, q') + \frac{dV}{dq} = 0$$

or, rewritten as a system of first-order ODEs,

$$q' = y$$
$$y' = -f(q, y) - dV/dq \tag{7}$$

where $-f(q, y)$ represents the frictional force; the function $f(q, y)$ always has the sign of y.

At low velocities, $f(q, y) = by$ is a reasonably good approximation of the friction due to air, but higher powers of y are necessary at higher velocities. This latter fact is why reducing the speed limit actually helps reduce gasoline usage—there is less drag at lower speeds. If friction were only a linear function of velocity, the effects of a higher speed would be cancelled by the distance being covered in a shorter time, and the system would expend the same amount of energy in either case. But if friction depends on the cube of velocity, for instance, you gain a lot by going more slowly. We will examine more elaborate friction laws when we study the pumping of a swing, but for now we will use viscous damping with $f = by$.

Example 4: Let's model the motion of a linearized pendulum with and without damping:

$$\theta' = y$$
$$y' = -10\theta - by \tag{8}$$

where $b = 0$ (no damping), or $b = 1$ (viscous damping). If there is no damping, then one conserved quantity is

$$K = \frac{1}{2}y^2 + 5\theta^2 \tag{9}$$

The left graph in Figure 10.2 displays the integral surface defined by formula (9). The surface is a bowl whose cross-sections $K = K_0$ are ellipses. Projecting the ellipses downward onto the θy-plane gives the trajectories of system (8) with $b = 0$.

Once damping is turned on, the integral K in formula (9) no longer is constant on a trajectory. But the integral concept still gives a good geometric picture of the behavior of a system under damping, because the value of K decreases along trajectories. This fact follows from the following computation (using system (7)):

$$\frac{d}{dt}\left(\frac{y^2}{2} + V(q)\right) = y\frac{dy}{dt} + \frac{dV}{dq}\frac{dq}{dt} = y\left(-f(q, y) - \frac{dV}{dq}\right) + \frac{dV}{dq}y$$
$$= -yf(q, y) \leq 0$$

where the final inequality follows from the fact that $f(q, y)$ has the sign of y. In particular, the value of K along a solution of system (8) decreases, and will either tend to a finite limit, which can only happen if the solution tends to an equilibrium of the system, or the value of K will tend to $-\infty$. If V is bounded from below (as happens for all our examples), the latter does not happen.

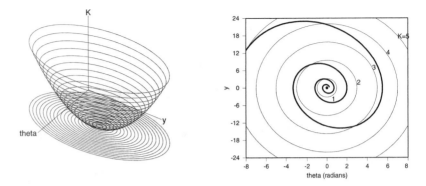

Figure 10.2: The left graph shows the integral surface $K = y^2/2 + 5\theta^2$ for the undamped, linearized system, $\theta' = y$, $y' = -10\theta$, and the projections of the level curves $K = K_0$. The right graph shows a trajectory of the damped, linearized system, $\theta' = y$, $y' = -10\theta - y$, as it cuts across the level curves of K, with K decreasing as it goes.

Example 5: Let's turn on viscous damping (take $b = 1$ in system (8)) and see what happens. The right side of Figure 10.2 shows a trajectory of the damped, linear pendulum system as it cuts across the level curves of the integral function $K = y^2/2 + 5\theta^2$. K decreases as the trajectory approaches the spiral sink at $\theta = 0$, $y = 0$. [The level curves of K are drawn by ODE Architect as trajectories of the undamped system (8) with $b = 0$.]

Now let's turn to the more realistic nonlinear pendulum and see how damping affects its motions.

Example 6: The nonlinear system is

$$\theta' = y$$
$$y' = -10\sin\theta - by \tag{10}$$

where $b = 0$ corresponds to no damping, and $b = 1$ gives viscous damping. In the no-damping case we can take the conserved quantity K to be

$$K = \frac{1}{2}y^2 - 10(\cos\theta - 1) \tag{11}$$

The left side of Figure 10.3 shows part of the surface defined by equation (11).

Example 7: With damping turned on (set $b = 1$ in system (10)) a trajectory with a high initial K-value may "swing over the pivot" several times before settling into a tightening spiral terminating at a sink, $\theta = 2n\pi$, $y = 0$, for some value of n. The right side of Figure 10.3 shows one of these trajectories as it swings over the pivot once, and then heads toward the point, $\theta = 2\pi$, $y = 0$, where $K = 0$.

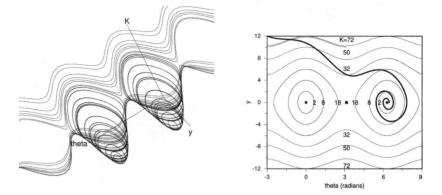

Figure 10.3: The left graph shows the surface $K = y^2/2 - 10(\cos\theta - 1)$ with two of its bowl-like projections that touch the θy-plane at equilibrium points of minimal K-value. The nonlinear pendulum system is $\theta' = y$, $y' = -10\sin\theta - by$ with $b = 0$. Turn on damping ($b = 1$) and watch a trajectory cut across the level sets $K = K_0$, with ever smaller values of K (right graph).

✓   Would you increase or decrease b to cause the trajectory starting at $\theta = -3$, $y = 12$ to approach $\theta = 0$, $y = 0$? How about $\theta = 10\pi$, $y = 0$? What would you need to do to get the trajectory to approach $\theta = -2\pi$, $y = 0$, or is this even possible?

◆ Separatrices

A trajectory is a *separatrix* if the long-term behavior of trajectories on one side is quite different from the behavior on the other side. As we saw in Chapters 6 and 7 each saddle comes equipped with four separatrices: two approach the saddle with increasing time (the *stable* separatrices for that saddle point) and two approach as time decreases (the *unstable* separatrices). These separatrices are of the utmost importance in understanding how solutions behave in the long term.

Example 8: The undamped system

$$\begin{aligned} \theta' &= y \\ y' &= -10\sin\theta \end{aligned} \tag{12}$$

has equilibrium points at $\theta = n\pi$, $y = 0$. According to the equilibrium calculations in ODE Architect, these points are centers if n is even, and saddles if n is odd. Each separatrix at a saddle enters (or leaves) the saddle tangent to an eigenvector of the Jacobian matrix evaluated at the point. ODE Architect gives us these eigenvectors after it has located the saddle.

Example 9: (*Plotting a Separatrix:*) To find a point approximately on a saddle separatrix, just take a point close to a saddle and on an eigenvector. Then solve

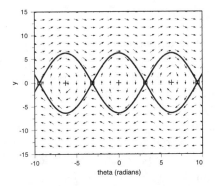

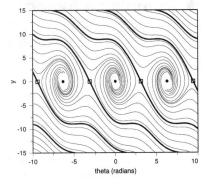

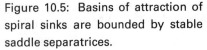

Figure 10.4: Saddle separatrices for the undamped, nonlinear pendulum system enclose centers.

Figure 10.5: Basins of attraction of spiral sinks are bounded by stable saddle separatrices.

forward and backward to obtain a reasonable approximation to a separatrix. For example, at $\theta = \pi$, $y = 0$, ODE Architect tells us that $(0.3015, 0.9535)$ is an eigenvector corresponding to the eigenvalue 3.162, and so the saddle separatrix is unstable. To graph the corresponding separatrix we choose as the initial point $\theta_0 = \pi + 0.003015$, $y = 0.009535$ which is in the direction of the eigenvector and very close to the saddle point. Figure 10.4 shows several separatrices of system (12). The squares indicate saddle points, and the plus signs inside the regions bounded by separatrices indicate centers.

✓ Describe the motions that correspond to trajectories inside the regions bounded by separatrices. Repeat with the region above the separatrices. Can a separatrix be both stable and unstable when considered as an entire trajectory?

Example 10: Add in viscous damping and the picture completely changes: Figure 10.5 shows the stable separatrices at the saddle points for the system

$$\theta' = y$$
$$y' = -10\sin\theta - y \tag{13}$$

The equilibrium points at $\theta = 2n\pi$, $y = 0$ are no longer centers, but attracting spiral points (the solid dots). The *basin of attraction* of each sink (i.e., the points on the trajectories attracted to the sink) is bounded by the four stable saddle separatrices.

✓ With a fine-tipped pen, draw the *unstable* separatrices at each saddle in Figure 10.5.

That's all we have to say about the motions of a constant-length pendulum for now. More (much more) is discussed in Chapter 12, where we add a driving term $F(t)$ to the pendulum equations.

◆ Pumping a Swing

Recall that in an *autonomous* differential equation, the time variable t does not appear explicitly. The central thing to realize is that *the ODE that models pumping a swing must be autonomous:* a child pumping the swing does not consult a watch when deciding how to lean back or sit up; the movements depend only on the position of the swing and its velocity. The swinger may change pumping strategies, deciding to go higher or slow down, but the modeling differential equation for any particular strategy should be autonomous, depending on various parameters which describe the strategy.

☞ If you use a different pumping strategy, make up a differential equation of your own!

If you observe a child pumping a swing, or do it yourself, you will find that one strategy is to lean back on the first half of the forward swing and to sit up the rest of the time. If you stand on the seat, the strategy is the same: you crouch during the forward down-swing, and stand up straight the rest of the time. The work is done when you bring yourself back upright during the forward up-swing, either by pulling on the ropes (if sitting), or simply by standing.

The pumping action effectively changes the length of the swing, which complicates the ODE considerably, for two reasons. Newton's second law must be stated differently, as will be shown below, and we must find an appropriate equation to model the changing length.

The question of friction is more subtle. Of course, the air creates a drag, but that is not the most important component of friction. We believe that things are quite different for a swing attached to the axle by something flexible, than if it were attached by rigid rods. Circus acrobats often drive swings right over the top; they always have rigid swings. We believe that a swing attached flexibly to the axle cannot be pumped to go over the top. Suppose the swing were to go beyond the horizontal—then at the end of the forward motion, the swinger would go into free-fall instead of swinging back; the jolt (heard as "ka-chunk") when the rope becomes tight again will drastically slow down the motion. If you get on a swing, you will find that this effect is felt before the amplitude of the swing reaches $\pi/2$; the ropes become loose near the top of the forward swing, and you slow down abruptly when they draw tight again.

We will now turn this description into a differential equation.

◆ Writing the Equations of Motion for Pumping a Swing

Modeling the pendulum with changing length requires a more careful look at Newton's second law of motion. The equation $F = ma = mq''$ is not correct when the mass is changing (as when you use a leaky bucket as the bob of a pendulum), or when the distance variable is changing with respect to position and velocity (as for the child on the swing). In cases such as this, force is the rate of change of the *momentum mq'*:

$$\text{Force} = (mq')' \tag{14}$$

When the mass and pendulum length are constant, equation (14) indeed reduces to the more familiar $F = ma$.

The analog in rotational mechanics about a pivot, where $q = L\theta$, is that the *torque* equals the rate of change of angular momentum:

$$\text{Torque} = (I\theta')'$$

where I is the moment of inertia (the rotational analog of the mass). If a force **F** is applied at a point p, then the torque about the pivot is the vector product $\mathbf{r} \times \mathbf{F}$, where **r** is the position vector from the pivot to p. For the undamped and nonlinear pendulum, the gravitational torque can be treated as the scalar $-mgL\sin\theta$, and the moment of inertia is $I = mL^2$. Then Newton's second law becomes

$$(mL^2\theta')' = -mgL\sin\theta \tag{15}$$

When L and m are constant, equation (15) is precisely the ODE of the undamped, nonlinear pendulum. In the case of the child pumping a swing, the mass m remains constant (and can be divided out of the equation), but L is *not* constant, so we must differentiate $L^2\theta'$ in equation (15) using the chain rule to get

$$2L\left(\frac{\partial L}{\partial \theta}\theta' + \frac{\partial L}{\partial \theta'}\theta''\right)\theta' + L^2\theta'' = -gL\sin\theta$$

or, in system form

$$\theta' = y$$
$$y' = -\frac{2y^2\partial L/\partial\theta + g\sin\theta}{2y\partial L/\partial y + L} \tag{16}$$

The person pumping the swing is changing L as a function of θ and y. For the reasons given in Screen 2.3 of Module 10 we will use the following formula for L:

$$L = L_0 + \frac{\Delta L}{\pi^2}\left(\frac{\pi}{2} - \arctan 10\theta\right)\left(\frac{\pi}{2} + \arctan 10y\right) \tag{17}$$

where L_0 is the distance from the axle to the center of gravity of the swinger when upright, and ΔL is the amount by which leaning back (or crouching) increases this distance. Note that

$$\frac{1}{\pi}\left(\frac{\pi}{2} - \arctan 10\theta\right)$$

is a smoothed-out step function: roughly 1 when $\theta < 0$ and 0 when $\theta > 0$. The jump from one value to the other is fairly rapid because of the factor 10; other values would be appropriate if you were to sit (or stand) up more or less suddenly. A similar analysis applies to the second arctan factor in formula (17).

As for friction with the swing, we will use

$$f(\theta, y) = \varepsilon y + \left(\frac{\theta}{1.4}\right)^6 y$$

The first term corresponds to some small viscous air resistance. Admittedly, the second term is quite ad-hoc, but it serves to describe some sort of insurmountable "brick wall," which somewhat suddenly takes effect when $\theta > 3/2 \sim \pi/2$. So it does seem to reflect our qualitative description.

Writing the differential equation as an autonomous system is now routine—an unpleasant routine since we need to differentiate L, which leads to pretty horrific formulas. But with this summary, we have tried to make the structure clear. Now let's get real and insert friction into modeling system (16):

$$\theta' = y$$

$$y' = -\frac{2y^2 \partial L/\partial\theta + g\sin\theta + \text{friction term}}{2y\partial L/\partial y + L} \tag{18}$$

where L is given by formula (17) and

$$\frac{\partial L}{\partial\theta} = -10\Delta L \frac{\frac{\pi}{2} + \arctan(10y)}{\pi^2(1 + 100\theta^2)}$$

$$\frac{\partial L}{\partial y} = 10\Delta L \frac{\frac{\pi}{2} - \arctan(10\theta)}{\pi^2(1 + 100y^2)} \tag{19}$$

$$\text{friction term} = \varepsilon y + \left(\frac{\theta}{1.4}\right)^6 y$$

Example 11: Now set $g = 32$, $L_0 = 4$, $\Delta L = 1$, and $\varepsilon = 0$ (no viscous damping), and use ODE Architect to solve system (18). Figure 10.6 shows that you can pump up a swing from rest at an initial angle of 0.25 radian (about 14°) within a reasonable time, but not from the tiny angle of 0.01 radian. Do you see the approach to a stable, periodic, high-amplitude oscillation? This corresponds to an attracting limit cycle in the θy-plane.

What happens if we put viscous damping back in? See for yourself by going to Screen 2.5 of Module 10 and clicking on several initial points in the θy-screen. You should see *two* limit cycles now:

- a large attracting limit cycle representing an oscillation of amplitude close to $\pi/2$, due to the "brick wall" friction term, and (for $\varepsilon > 0$)

- a small repelling limit cycle near the downward equilibrium, due to friction and viscous air resistance.

In order to get going, the child must move the swing outside the small limit cycle, either by cajoling someone into pushing her, or backing up with her feet on the ground. Once outside the small limit cycle, the pumping will push the trajectory to the attracting limit cycle, where it will stay until the child decides to slow down.

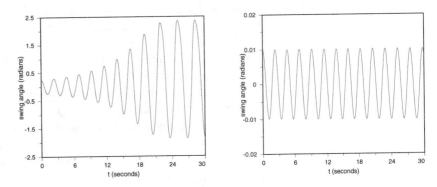

Figure 10.6: Successful pumping (left graph) starts at a moderately high an-
gle ($\theta = 0.25$ radian). If θ_0 is small (e.g., $\theta_0 = 0.01$ rad), then pumping doesn't
help much (right graph).

Please note that this structure of the phase plane, with two limit cycles, is
necessary in order to account for the observed behavior: the origin must be a
sink because of air resistance, and you cannot have an attracting limit cycle
surrounding a sink without another limit cycle in between.

✓ Does the system without viscous damping have a small repelling limit
cycle?

◆ Geodesics

Geodesics on a surface are curves that minimize length between sufficiently
close points on the surface; they may, but need not, minimize length between
distant points.

Example 12: Straight lines are geodesics on planes, and they minimize the
distance between arbitrary points. Great circles are geodesics on the unit
sphere, but they only minimize length between pairs of points if you travel
in the right direction. If you travel along the equator your path will be shortest until you get half-way around the world; but further along, you would have
done better to go the other way.

To look for geodesics, we use the fact that parametrization of a curve γ
by its arc length s results in traversing a curve at constant speed 1, that is,
$|d\gamma/ds|$ is always 1.

On a surface in three-dimensional space, a geodesic γ is a curve for which
the vector $d^2\gamma/ds^2$ is perpendicular to the surface at the point $\gamma(s)$. For now,
let's assume that all curves are parametrized by arc length, so γ' means $d\gamma/ds$.

If any curve γ (not necessarily a geodesic) on the surface is parametrized
at constant speed, we are guaranteed that γ'' is perpendicular to γ', but not
necessarily to the surface. To see this, observe that $\gamma' \cdot \gamma' = 1$, where γ' is

the velocity vector for the curve γ and the "dot" indicates the *dot* (or *scalar*) *product* of two vectors. Differentiating the dot product equation, we have $\gamma'' \cdot \gamma' + \gamma' \cdot \gamma'' = 0$, so γ'' is perpendicular to γ' (or else is the zero vector).

The statement that γ'' is perpendicular to the *surface* says that γ is going as "straight" as it can in the surface, and that the surface is exerting no force which would make the curve bend away from its path. Such a curve is a geodesic. See the book by Do Carmo for a full explanation of why geodesics defined as above minimize the distance between nearby points.

Example 13: On a sphere, the parallels of latitude yield acceleration vectors in the plane of the parallel and perpendicular to the parallel (but not in general perpendicular to the surface), whereas any great circle yields acceleration vectors pointing toward the center of the sphere and hence perpendicular to both the great circle and to the surface. The great circles are geodesics, but the parallels (except for the equator) are not.

◆ Geodesics on a Surface of Revolution

Suppose that

$$x = f(u), \quad z = g(u)$$

is a parametrization by arc length u of a curve in the xz-plane. One consequence of this parametrization is that $(f'(u))^2 + (g'(u))^2 = 1$. Let's rotate the curve by an angle θ around the z-axis, to find the surface parametrized by

$$P(u, \theta) = \begin{bmatrix} f(u)\cos\theta \\ f(u)\sin\theta \\ g(u) \end{bmatrix}$$

Let's suppose that curves γ on the surface are parametrized by arc length s and, hence, these curves have the parametric equation

$$\gamma(s) = \begin{bmatrix} f(u(s))\cos\theta(s) \\ f(u(s))\sin\theta(s) \\ g(u(s)) \end{bmatrix}$$

and we need to differentiate this twice to find

$$\gamma'(s) = \begin{bmatrix} f'(u(s))u'(s)\cos\theta(s) - f(u(s))\sin\theta(s)\theta'(s) \\ f'(u(s))u'(s)\sin\theta(s) + f(u(s))\cos\theta(s)\theta'(s) \\ g'(u(s))u'(s) \end{bmatrix}$$

$$\gamma''(s) = u'' \begin{bmatrix} f'(u)\cos\theta \\ f'(u)\sin\theta \\ g'(u) \end{bmatrix} + (u')^2 \begin{bmatrix} f''(u)\cos\theta \\ f''(u)\sin\theta \\ g''(u) \end{bmatrix} + 2u'\theta' \begin{bmatrix} -f'(u)\sin\theta \\ f'(u)\cos\theta \\ 0 \end{bmatrix}$$

$$- (\theta')^2 \begin{bmatrix} f(u)\cos\theta \\ f(u)\sin\theta \\ 0 \end{bmatrix} + \theta'' \begin{bmatrix} -f(u)\sin\theta \\ f(u)\cos\theta \\ 0 \end{bmatrix} \quad (20)$$

This array is pretty terrifying, but the two equations

$$\gamma'' \cdot \frac{\partial P}{\partial u} = 0, \quad \text{and} \quad \gamma'' \cdot \frac{\partial P}{\partial \theta} = 0 \tag{21}$$

which express the fact that γ'' is perpendicular to the surface, give

$$u'' - (\theta')^2 f(u) f'(u) = 0 \quad \text{and} \quad 2u' f(u) f'(u)\theta' + \theta''(f(u))^2 = 0 \tag{22}$$

That γ'' is perpendicular to the surface if formulas (21) hold follows because the vectors $\partial P/\partial u$ and $\partial P/\partial \theta$ span the tangent plane to the surface at the point (θ, u). Formulas (22) follow from formulas (21), from the fact that $(f'(u))^2 + (g'(u))^2 = 1$, and from the formulas

$$\frac{\partial P}{\partial u} = \begin{bmatrix} f'(u)\cos\theta \\ f'(u)\sin\theta \\ g'(u) \end{bmatrix}, \qquad \frac{\partial P}{\partial \theta} = \begin{bmatrix} -f(u)\sin\theta \\ f(u)\cos\theta \\ 0 \end{bmatrix}$$

The quantity

$$M = (f(u))^2\theta' \tag{23}$$

is conserved along a trajectory of system (22) since

$$\frac{d}{ds}[(f(u))^2\theta'] = (f(u))^2\theta'' + 2u'\theta' f' f = 0$$

The integral M behaves like *angular momentum* (see Exploration 10.5 for the central force context that first gave rise to this notion).

Substituting $\theta' = M/(f(u))^2$ into the first ODE of equations (22) gives

$$u'' - \frac{M^2 f'(u)}{(f(u))^3} = 0 \tag{24}$$

Using equations (23) and (24), we obtain a system of ODEs for the geodesics on a surface of revolution:

$$\theta' = \frac{M}{(f(u))^2}$$
$$u' = w \tag{25}$$
$$w' = \frac{M^2 f'(u)}{(f(u))^3}$$

We recognize that ODE (24) is of the form of ODE (4), so

$$u'' = -\frac{d}{du}\frac{M^2}{2(f(u))^2}$$

and we can analyze this ODE by the phase plane and conservation methods used earlier. Let us now specialize to the torus.

◆ Geodesics on a Torus

Rotate a circle of radius r about a line lying in the plane of the circle to obtain a torus. If R is the distance from the line to the center of the circle, then in the above equations we can set

$$x = f(u) = R + r\cos u$$
$$z = g(u) = r\sin u$$

If we set $r = 1$, then we have $(f')^2 + (g')^2 = 1$ as required in the derivation of the geodesic ODEs. The system of geodesic ODEs (25) becomes

$$\theta' = \frac{M}{(R + \cos u)^2}$$
$$u' = w \tag{26}$$
$$w' = -\frac{M^2 \sin u}{(R + \cos u)^3}$$

where M is a constant. The variable u measures the angle up from the outer equator of the torus, and θ measures the angle around the outer equator from some fixed point. Figure 10.7 shows seventeen geodesics through the point $\theta_0 = 0$, $u_0 = 0$ with w_0 sweeping from -8 to 8. In Figure 10.7 and subsequent figures we take $R = 3$ and $M = 16$. Figure 10.8 shows the geodesic curves in the θu-plane (left graph) and in the uu'-plane (right). Note the four outlying geodesics that coil around the torus, repeatedly cutting both the outer $[u = 2n\pi]$ and the inner $[u = (2n + 1)\pi]$ equators and periodically going through the hole of the donut. Twelve geodesics oscillate about the geodesic along the outer equator.

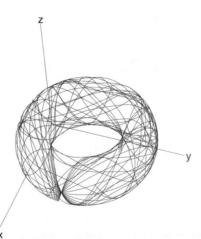

Figure 10.7: Seventeen geodesics through a point on the outer equator.

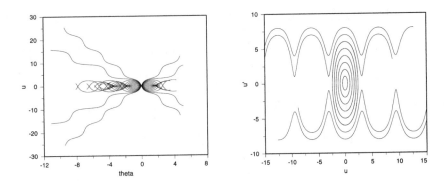

Figure 10.8: The seventeen geodesics of Figure 10.7 drawn in the θu-plane (left) and in the uu'-plane (right).

Figure 10.9 shows the outer and inner equatorial geodesics (the horizontal lines) in the θu-plane, as well as three curving geodesics starting at $\theta_0 = 0$, $u_0 = 0$. One oscillates about the outer equator six times in one revolution (i.e., as θ increases from 0 to 2π). The other two start with values of y_0 that take them up over the torus and near the inner equator. One of these geodesics turns back and slowly oscillates about the outer equator. The other starts with a slightly larger value of y_0, cuts across the inner equator, and slowly coils around the torus. This suggests that the inner equator ($u = \pi$) is a *separatrix* geodesic, dividing the geodesics into those that oscillate about the outer equator from those that coil around the torus. This separatrix is

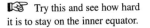 Try this and see how hard it is to stay on the inner equator.

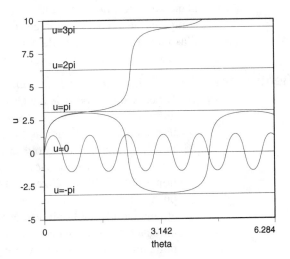

Figure 10.9: Equatorial geodesics (lines), a geodesic that rapidly oscillates around the outer equator, another that oscillates slowly around the outer equator, and a third that slowly coils around the torus.

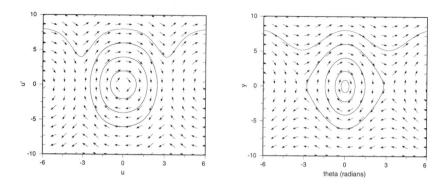

Figure 10.10: The graphs of the toroidal geodesics in the uu'-plane (left) look like the trajectories of an undamped, nonlinear pendulum (right).

unstable in the sense that if you start a geodesic near the inner equator (say at $\theta_0 = 3.14$, $y_0 = 0$) and solve the system (26), then the geodesic moves away from the separatrix.

Why do we call this geodesic model a "friend of the pendulum"? Take a look at the u' and w' ODEs in system (26). Note that if we delete the term "$\cos u$" from the denominator of the w' equation, then we obtain the system

$$u' = w$$

$$w' = -\frac{M^2}{R^3}\sin u \tag{27}$$

which is precisely the system for an undamped, nonlinear pendulum with $g/L = M^2/R^3$. This fact suggests that geodesics of (26) plotted in the uu'-plane will look like trajectories of the pendulum system (27). Figure 10.10 compares the two sets of trajectories and shows how much alike they are. This illustrates a general principle (which, like most principles, has its exceptions): *If two systems of ODEs resemble one another, so will their trajectories.*

References   Arnold, V.I., *Ordinary Differential Equations* (1973: M.I.T.)

Note: Arnold's book is the classical text. Much of the considerable literature on modeling swings has been influenced by his description, which uses a nonautonomous length function, $L(t)$, instead of $L(\theta, \theta')$. In Sections 27.1 and 27.6, Arnold shows that if the child makes pumping motions at the right frequency at the bottom position of the swing, the motion eventually destabilizes, and the swing will start swinging without any push. This is correct (we have seen it done, though it requires pumping at an uncomfortably high frequency), but appears to us to be completely unrelated to how swings are actually pumped. Since the usual swing-pumping is done without reference to a clock, a proper model must certainly give an autonomous equation.

Do Carmo, M.P., *Differential Geometry of Curves and Surfaces* (1976: Prentice-Hall)

Halliday, D., Resnick, R., *Physics*, (v. I and II), 3rd ed. (1977-78: John Wiley & Sons, Inc.)

Hubbard, J.H., and West, B.H., *Differential Equations: A Dynamical Systems Approach, Part II: Higher Dimensional Systems*, (1995: Springer-Verlag) especially Ch. 6.5 and 8.1, which have more on conservation of energy.

Pennington, R., "The Pendulum Revisited (with MAPLE)" in *C·ODE·E*, Spring/Summer 1996, www.math.hmc.edu/codee.

Stewart, I., *Does God Play Dice?* The Mathematics of Chaos (1989: Blackwell Publishers) Ch. 5 The One Way Pendulum.

Werkes, S., Rand, R., Ruina, A., "Modeling the Pumping of a Swing" in *C·ODE·E*, Winter/Spring 1997, www.math.hmc.edu/codee.

Answer questions in the space provided, or on
attached sheets with carefully labeled graphs. A
notepad report using the Architect is OK, too.

Name/Date _____

Course/Section _____

Exploration 10.1. Explorations of Basic Pendulum Equation

1. If the nonlinear pendulum ODE (3c) is approximated by the linear ODE (3a),
    how closely do the trajectories and the component curves of the two ODEs
    match up? Screen 1.2 in Module 10 will be a big help here.

2. What would motions of the system, $x' = y$, $y' = -V(x)$, look like under
    different potential functions, such as $V(x) = x^4 - x^2$? What happens if a
    viscous damping term $-y$ is added to the second ODE of the system? Use
    graphical images like those in Figures 10.2 and 10.3 to guide your analysis.
    Use ODE Architect to draw trajectories in the xy-plane for both the undamped
    and damped case. Identify the equilibrium points in each case as saddles,
    centers, sinks, or sources. Plot the stable and the unstable saddle separatrices
    (if there are any) and identify the basin of attraction of each sink. [*Suggestion:*
    Use the Equilibrium feature of ODE Architect to locate the equilibrium points,
    calculate Jacobian matrices, find eigenvalues and eigenvectors, and so help to
    determine the nature of those points.]

3. Find all solutions of the undamped and linearized pendulum ODE,

$$\theta'' + (g/L)\theta = 0$$

Show that all solutions are periodic of period $2\pi\sqrt{L/g}$. Are all solutions of the corresponding nonlinear pendulum ODE, $\theta'' + (g/L)\sin\theta = 0$, periodic? If the latter ODE has periodic solutions, compare the periods with those of solutions of the linearized ODE that have the same initial conditions.

4. Use the sweep and the animate features of ODE Architect to make "movies" of the solution curves and the trajectories of the nonlinear pendulum ODE, $\theta'' + b\theta + \sin\theta = 0$, where $\theta_0 = 0$, $\theta'_0 = 10$, and b is a nonnegative parameter. Interpret what you see in terms of the motions of a pendulum. In this regard, you may want to use the model-based pendulum animation feature of ODE Architect.

Answer questions in the space provided, or on
attached sheets with carefully labeled graphs. A
notepad report using the Architect is OK, too.

Name/Date _____

Course/Section _____

Exploration 10.2. Physical Variations for Child on a Swing

1. Module 10 and the text of this chapter describe a swing-pumping strategy
    where the swinger changes position only on the first half of the forward swing
    (i.e., where θ is negative but θ' is positive). Is this the strategy you would use
    to pump a swing? Try pumping a swing and then describe in words your most
    successful strategy.

2. Rebuild the model for the length function $L(\theta, \theta')$ of the "swing pendulum" to model your own pumping scenario. [*Suggestion:* Change the arguments of the arctan function used in Module 10 and the text of this chapter.] Use the ODE Architect to solve your set of ODEs. From plots of $t\theta$-curves and of $\theta\theta'$-trajectories, what do you conclude about the success of your modeling and your pumping strategy?

Answer questions in the space provided, or on
attached sheets with carefully labeled graphs. A
notepad report using the Architect is OK, too.

Name/Date _____

Course/Section _____

Exploration 10.3. Bifurcations

In these problems you will study the bifurcations in the swing-pumping model
of Module 10 and this chapter as the viscous damping constant ε or the in-
cremental pendulum length ΔL is changed.

1. There is a Hopf bifurcation for the small-amplitude repelling limit cycle at
$\varepsilon = 0$ for the swing-pumping system (18) and (19). Plot lots of trajectories
near the origin $\theta = 0$, $y = 0$ for values of ε above and below $\varepsilon = 0$ and
describe what you see. What does the ODE Architect equilibrium feature tell
you about the nature of the equilibrium point at the origin if $\varepsilon < 0$? If $\varepsilon = 0$?
If $\varepsilon > 0$?

2. Now sweep ΔL through a series of values and watch what happens to the large-amplitude attracting limit cycle. At a certain value of ΔL you will see a sudden change (called a *homoclinic, saddle-connection bifurcation*). What is this value of ΔL? Plot lots of trajectories for various values of ΔL and describe what you see.

Answer questions in the space provided, or on
attached sheets with carefully labeled graphs. A
notepad report using the Architect is OK, too.

Name/Date _____

Course/Section _____

Exploration 10.4. Geodesics on a Torus

The basic initial value problem for a geodesic starting on the outer equator of a torus is

$$\theta' = \frac{M}{(R + \cos u)^2}$$

$$u'' = -\frac{M^2 \sin u}{(R + \cos u)^3} \tag{28}$$

$$u(0) = 0, \quad u'(0) = \alpha, \quad \theta(0) = 0$$

where M is a constant.

1. Make up your own "pretty pictures" of geodesic sprays on the surface of the torus by varying $u'(0)$. Explain what each geodesic is doing on the torus. If two geodesics through $u_0 = 0$, $\theta_0 = 0$ intersect at another point, which provides the shortest path between two points? Is every "meridian", $\theta = $ const., a geodesic? Is every "parallel", $u = $ const., a geodesic?

2. Repeat Problem 1 at other initial points on the torus, including a point on the inner equator.

3. Explore different values for R (between 2 and 5) for the torus—what does it mean for the solutions of the ODEs for the geodesics? To what extent does the ugly denominator in the ODEs mess up the similarity to the nonlinear pendulum equation?

    Answer by discussing effects on θu-phase portraits.

Answer questions in the space provided, or on attached sheets with carefully labeled graphs. A notepad report using the Architect is OK, too.

Name/Date _____

Course/Section _____

Exploration 10.5. The Central Force and Kepler's Laws

An object at position $\mathbf{r}(t)$ (relative to a fixed coordinate frame) is moving under a central force if the force points toward or away from the origin, with a magnitude which depends only on the distance r from the origin. This is modeled by the differential equation $\mathbf{r}'' = f(r)\mathbf{r}$, where we will take $\mathbf{r}(t)$ to be a vector moving in a fixed plane.

Example 14: (*Newton's law of gravitation*) This, as applied to a planet and the sun, is perhaps the most famous differential equation of all of science. Newton's law describes the position of the planet by the differential equation

$$\mathbf{r}'' = -\frac{AG}{r^3}\mathbf{r},$$

where $\mathbf{r}$ is the vector from the center of gravity of the two bodies (located, for all practical purposes, at the sun) to the planet, G is the universal gravitational constant, and $A = M^3/(m+M)^2$, where m is the mass of the planet (so for all practical purposes, A is the mass of the sun).

1. Newton's law of gravitation is often called the "inverse square law," not the "inverse cube law." Explain.

2. The way to analyze a central force problem is to write it in polar coordinates, where

☞ Another way to write the vector $\mathbf{r}$ is $\mathbf{r} = \hat{\mathbf{i}}\cos\theta + \hat{\mathbf{j}}\sin\theta$, where $\hat{\mathbf{i}}$ and $\hat{\mathbf{j}}$ are unit vectors along the positive x- and y-axes, respectively.

$$\mathbf{r} = r[\cos\theta, \sin\theta]$$
$$\mathbf{r}' = r'[\cos\theta, \sin\theta] + r\theta'[-\sin\theta, \cos\theta]$$
$$\mathbf{r}'' = (r'' - r(\theta')^2)[\cos\theta, \sin\theta] + (2r'\theta' + r\theta'')[-\sin\theta, \cos\theta]$$

Show that the central force equation $\mathbf{r}'' = f(r)\mathbf{r}$ yields

$$2r'\theta' + r\theta'' = 0 \tag{29}$$

and

$$r'' - r(\theta')^2 = rf(r) \tag{30}$$

3. Show that the quantity $M = r^2\theta'$ is constant as a function of time during a motion in a central force system, using equation (29).

   The quantity M (now called the angular momentum of the motion) was singled out centuries ago as a quantity of interest precisely because of the derivation above. You should see that the constancy of M is equivalent to *Kepler's second law*: the vector **r** sweeps out equal areas in equal times.

4. Substitute $\theta' = M/r^2$ into equation (30) and show that, for each value of the particular central force $f(r)$ and each angular momentum M, the resulting differential equation is of the expected form.

5. Specialize to Newton's inverse square law with $k = AG$ and show that the resulting system becomes

$$r'' = -\frac{k}{r^2} + \frac{M^2}{r^3}$$

or the system

$$r' = y$$

$$y' = -\frac{k}{r^2} + \frac{M^2}{r^3}$$

Make a drawing of the phase plane for this system, and analyze this drawing using the conserved quantity K, where

$$K(r, y) = \frac{y^2}{2} + \frac{k}{r} - \frac{M^2}{2r^2}$$

K is evidently defined only for $r > 0$, and K has a unique minimum, so the level curves of K are simple closed curves for $K \ll 0$, corresponding to the elliptic orbits of *Kepler's first law*, an unbounded level curve when $K = 0$, corresponding to a parabolic orbit, and other unbounded curves for $K > 0$ which correspond to hyperbolic orbits. (For discussion of these three cases and their relation to conic sections, see Hubbard and West, Part II, Section 6.7 pp. 43–47.)

11

Applications of
Series Solutions

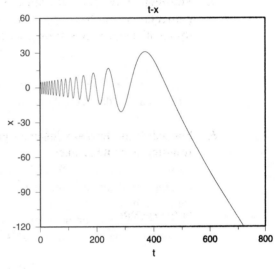

t-x

An aging spring stretches.

Overview   Many phenomena, especially those explained by Newton's Second Law, can be modeled by second-order linear ODEs with variable coefficients, for example:

1. Robot arms, which are modeled by a spring-mass equation with a time-varying damping coefficient; and

2. Aging springs, which are modeled by a spring-mass equation with a time-varying spring constant.

These two applications illustrate very different ways in which series solutions can be used to solve linear ODEs with nonconstant coefficients.

Key words   Infinite series; recurrence formula; ordinary point; singular point; regular singular point; Bessel's equations; Bessel functions; aging spring; lengthening pendulum

See also   Chapter 4 for second-order linear ODEs with constant coefficients (i.e., without the time-dependence).

◆ Infinite Series

Certain second-order linear ODEs with nonconstant coefficients have been studied extensively, so their properties are well-known. We will look at some of these ODEs in the chapter.

If the general linear homogeneous (undriven) second-order ODE

$$x'' + p(t)x' + q(t)x = 0 \tag{1}$$

has coefficients p and q that are not both constants, the methods of Chapter 4 don't work. However, sometimes we can write a solution $x(t)$ as a power series:

$$x(t) = \sum_{n=0}^{\infty} a_n(t - t_0)^n \tag{2}$$

where we use ODE (1) to determine the coefficients a_n. Much useful information can be deduced about an ODE when its solutions can be expressed as power series.

☞ Look in your calculus book for Taylor series. The term "analytic" is frequently used for functions with convergent Taylor Series.

If a function $x(t)$ has a convergent Taylor series $x(t) = \sum a_n(t - t_0)^n$ in some interval about $t = t_0$, then $x(t)$ is said to be *analytic* at t_0. Since all derivatives of analytic functions exist, the derivatives x' and x'' of x can be obtained by differentiating that series term by term, producing series with the same radius of convergence as the series for x. If we substitute these series into ODE (1), we can determine the coefficients a_n. To begin with, a_0 and a_1 are equal to the initial values $x(t_0)$ and $x'(t_0)$, respectively.

✓ "Check" your understanding by evaluating the series (2) at $t = t_0$ to show that $a_0 = x(t_0)$. Now differentiate series (2) term by term to obtain a series for $x'(t)$; evaluate this series at t_0 to find that $a_1 = x'(t_0)$. Does a_2 equal $x''(t_0)$?

◆ Recurrence Formulas

A *recurrence formula* for the coefficients a_n is a formula that defines each a_n in terms of the coefficients $a_0, a_1, \ldots, a_{n-1}$. To find such a formula, we have to express each of the terms in ODE (1) [i.e., x'', $p(t)x'$, and $q(t)x$] as power series about $t = t_0$, which is the point at which the initial conditions are given. Then we combine these series to obtain a single power series which, according to ODE (1), must sum to zero for all t near t_0. This implies that the coefficient of each power of $t - t_0$ must be equal to zero, which yields an equation for each a_n in terms of the preceding coefficients $a_0, a_1, \ldots, a_{n-1}$.

Example: Finding a recurrence formula

☞ We chose a first-order ODE for simplicity.

Let's solve the first-order IVP $x' + tx = 0$, $x(0) = 1$. First we write $x(t)$ in the form

$$x(t) = \sum_{n=0}^{\infty} a_n t^n$$

where we have chosen $t_0 = 0$. The derivative of $x(t)$ is then

$$x'(t) = \sum_{n=1}^{\infty} n a_n t^{n-1}$$

Substituting this into the given ODE, we get

$$x' + tx = \sum_{n=1}^{\infty} n a_n t^{n-1} + \sum_{n=0}^{\infty} a_n t^{n+1} = 0$$

To make the power of t the same in both sums, replace n by $n - 2$ in the second sum to obtain

☞ Notice in the second summation that n starts at 2, rather than 0. Do you see why?

$$\sum_{n=1}^{\infty} n a_n t^{n-1} + \sum_{n=2}^{\infty} a_{n-2} t^{n-1} = a_1 + \sum_{n=2}^{\infty} [n a_n + a_{n-2}] t^{n-1} = 0$$

The last equality is true if and only if $a_1 = 0$ and, if for every $n \geq 2$, we have that $n a_n + a_{n-2} = 0$. Therefore, the desired recurrence formula is

$$a_n = \frac{-a_{n-2}}{n}, \quad n = 2, 3, \ldots \tag{3}$$

Since $a_1 = 0$, formula (3) shows that the coefficients $a_3, a_5, \ldots, a_{2k+1}, \ldots$ must all be zero; and $a_2 = -a_0/2$, $a_4 = -a_2/4 = a_0/(2 \cdot 4)$, $\ldots$. With a little algebra you can show that the series for $x(t)$ is

$$x(t) = a_0 - \frac{a_0}{2} t^2 + \frac{a_0}{2 \cdot 4} t^4 - \frac{a_0}{2 \cdot 4 \cdot 6} t^6 + \cdots$$

which can be simplified to

$$x(t) = a_0 \left(1 - \frac{t^2}{2} + \frac{1}{2!} \left(\frac{t^2}{2} \right)^2 - \frac{1}{3!} \left(\frac{t^2}{2} \right)^3 + \cdots \right)$$

If the initial condition $a_0 = x(0) = 1$ is used, this becomes the Taylor Series for $e^{-t^2/2}$ about $t_0 = 0$. Although the series solution to the IVP, $x' + x = 0$, $x(0) = 1$, can be written in the form of a familiar function, for most IVPs that is rarely possible and usually the only form we can obtain is the series form of the solution.

✓ Check that $x(t) = e^{-t^2/2}$ is a solution of the IVP $x' + tx = 0$, $x(0) = 1$.

◆ Ordinary Points

If $p(t)$ and $q(t)$ are both analytic at t_0, then t_0 is called an *ordinary point* for the differential equation $x''(t) + p(t)x'(t) + q(t)x(t) = 0$. At an ordinary point, the method illustrated in the preceding example always produces solutions written in series form. The following theorem states this more precisely.

Ordinary Points Theorem. If t_0 is an ordinary point of the second-order differential equation

$$x'' + p(t)x' + q(t)x = 0 \qquad (4)$$

that is, if $p(t)$ and $q(t)$ are both analytic at t_0, then the general solution of ODE (4) is given by the series

$$x(t) = \sum_{n=0}^{\infty} a_n(t - t_0)^n = a_0 x_1(t) + a_1 x_2(t) \qquad (5)$$

where a_0 and a_1 are arbitrary and, for each $n \geq 2$, a_n can be written in terms of a_0 and a_1. When this is done, we get the right-hand term in formula (5), where $x_1(t)$ and $x_2(t)$ are linearly independent solutions of ODE (4) that are analytic at t_0. Further, the radius of convergence for each of the series solutions $x_1(t)$ and $x_2(t)$ is at least as large as the smaller of the two radii of convergence for the series for $p(t)$ and $q(t)$.

One goal of Module 11 is to give you a feeling for the interplay between infinite series and the functions they represent. In the first submodule, the position $x(t)$ of a robot arm is modeled by the second-order linear ODE

☞ Note that $p(t) = Ct$ and $q(t) = k$ are analytic for all t.

$$x'' + Ctx' + kx = 0 \qquad (6)$$

where C and k are positive constants. Using the methods of the earlier example, we can derive a series solution (with $t_0 = 0$)

$$x(t) = 1 - \frac{kt^2}{2!} + \frac{k(2C+k)t^4}{4!} - \frac{k(2C+k)(4C+k)t^6}{6!} + \cdots \qquad (7)$$

that satisfies $x(0) = 1$, $x'(0) = 0$. We then have to to determine how quickly the arm can be driven from the position $x = 1$ to $x = 0.005$ without letting x go below zero. The value of k is fixed at 9, so that only C is free to vary. When $C = k$, it turns out that series (7) is the Taylor series for $e^{-kt^2/2}$ about $t = 0$. It can then be demonstrated numerically, using ODE Architect, that $C = 9$ produces a solution that stays positive and is an optimal solution in the sense of requiring the least time for the value of x to drop from 1 to 0.005.

☞ Historically, new functions in engineering, science, and mathematics have often been introduced in the form of series solutions of ODEs.

In the majority of cases, however, it is *not* possible to recognize the series solution as one of the standard functions of calculus. Then the only way to approximate $x(t)$ at a given value of t is by summing a large number of terms

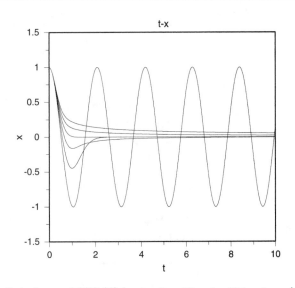

Figure 11.1: Solutions of ODE (6) for $k = 9$, $x(0) = 1$, $x'(0) = 0$, and $C = 0, 3, 6,$ 9, 12, 15. Which is the $C = 15$ curve?.

in the series, or by using a numerical solver to solve the corresponding IVP. ODE Architect was used to graph solutions of ODE (6) for several values of C (Figure 11.1).

What if t_0 is *not* an ordinary point for the ODE, $x'' + p(t)x' + q(t)x = 0$, that is, what if $p(t)$ or $q(t)$ is not analytic at t_0? For example, in the ODE $x'' + x/(t-1) = 0$, $q(t)$ is not analytic at $t_0 = 1$. Such a point is said to be a *singular point* of the ODE. For example, $t_0 = 1$ is a singular point for the ODE $x'' + x/(t-1) = 0$. Next we show how to deal with ODEs with certain kinds of singular points.

✓   Is $t = 0$ an ordinary point or a singular point of $x'' + t^2 x = 0$? What about $x'' + (\sin t)x = 0$ and $x'' + x/t = 0$?

◆ Regular Singular Points

A singular point of the ODE $x''(t) + p(t)x'(t) + q(t)x(t) = 0$ is a *regular singular point* if both $(t - t_0)p(t)$ and $(t - t_0)^2 q(t)$ are analytic at t_0. In this case we'll have to modify the method to find a series solution to the ODE.

✓   Is $t = 0$ a regular singular point of $x'' + x'/t + x = 0$?   What about $x'' + x' + x/t^2 = 0$ and $x'' + x' + x/t^3 = 0$?

Since $(t - t_0)p(t)$ and $(t - t_0)^2 q(t)$ are analytic at t_0, they have power series expansions centered at t_0:

$$(t - t_0)p(t) = P_0 + P_1(t - t_0) + P_2(t - t_0)^2 + \cdots$$
$$(t - t_0)^2 q(t) = Q_0 + Q_1(t - t_0) + Q_2(t - t_0)^2 + \cdots$$

As we shall soon see, the constant coefficients, P_0 and Q_0, in these two series are particularly important. The roots of the quadratic equation (called the *indicial equation*)

$$r(r - 1) + P_0 r + Q_0 = 0 \tag{8}$$

☞ Assume that the roots of the indicial equation are real numbers.

are used in solution formula (9) below.

A theorem due to Frobenius tells us how to modify our original method of constructing power series solutions so that we can obtain series solutions near regular singular points.

Frobenius' Theorem. If t_0 is a regular singular point of the second-order differential equation $x''(t) + p(t)x'(t) + q(t)x(t) = 0$, then there is at least one series solution at t_0 of the form

☞ The second summation is called the *Frobenius series*.

$$x_1(t) = (t - t_0)^{r_1} \sum_{n=0}^{\infty} a_n(t - t_0)^n = \sum_{n=0}^{\infty} a_n(t - t_0)^{n+r_1} \tag{9}$$

where r_1 is the larger of the two roots r_1 and r_2 of the indicial equation.

☞ Consult the references for detailed instructions on how to find the coefficients a_n.

The coefficients a_n can be determined in the same way as in the earlier example: differentiate twice, substitute the series for qx_1, px_1', and x_1'' into the given differential equation, and then find a recurrence formula.

Here are a few things to keep in mind when finding a Frobenius series.

1. The roots of the indicial equation may not be integers, in which case the series representation of the solution would not be a power series, but is still a valid series.

2. If $r_1 - r_2$ is not an integer, then the smaller root r_2 of the indicial equation generates a second solution of the form

$$x_2(t) = (t - t_0)^{r_2} \sum_{n=0}^{\infty} b_n(t - t_0)^n$$

which is linearly independent of the first solution $x_1(t)$.

3. When $r_1 - r_2$ is an integer, a second solution of the form

$$x_2(t) = Cx_1(t) \ln(t - t_0) + \sum_{n=0}^{\infty} b_n(t - t_0)^{n+r_2}$$

exists, where the values of the coefficients b_n are determined by finding a recurrence formula, and C is a constant. The solution $x_2(t)$ is linearly independent of $x_1(t)$.

◆ Bessel Functions

For any nonnegative constant p, the differential equation

$$t^2 x''(t) + t x'(t) + (t^2 - p^2) x(t) = 0$$

is known as *Bessel's equation of order p*, and its solutions are the *Bessel functions of order p*. In normalized form, Bessel's equation becomes

$$x''(t) + \frac{1}{t} x'(t) + \left(\frac{t^2 - p^2}{t^2} \right) x(t) = 0$$

☞ If t is very large, Bessel's equation looks like the harmonic oscillator equation, $x'' + x = 0$.

From this we can see that $tp(t) = 1$ and $t^2 q(t) = t^2 - p^2$, so that $tp(t)$ and $t^2 q(t)$ are analytic at $t_0 = 0$. Therefore zero is a regular singular point and, using equation (8), we find that the indicial equation is

☞ The roots of the indicial equation are are p and $-p$.

$$r(r - 1) + r - p^2 = r^2 - p^2 = 0$$

Application of Frobenius' Theorem yields a solution J_p given by the formula

☞ Consult the references for the derivation of the formula for $J_p(t)$.

$$J_p(t) = t^p \sum_{n=0}^{\infty} \frac{(-1)^n}{2^{2n} n! (p+1)(p+2) \cdots (p+n)} t^{2n}$$

The function $J_p(t)$ is called the *Bessel function of order p of the first kind*. The series converges and is bounded for all t. If p is not an integer, it can be shown that a second solution of Bessel's equation is $J_{-p}(t)$ and that the general solution of Bessel's equation is a linear combination of $J_p(t)$ and $J_{-p}(t)$.

For the special case $p = 0$, we get the function $J_0(t)$ used in the aging spring model in the second submodule of Module 11:

$$J_0(t) = \sum_{n=0}^{\infty} \frac{(-1)^n}{(n!)^2} \left(\frac{t}{2} \right)^{2n} = 1 - \frac{t^2}{4} + \frac{t^4}{64} - \frac{t^6}{2304} + \cdots$$

Note that even though $t = 0$ is a singular point of the Bessel equation of order zero, the value of $J_0(0)$ is finite $[J_0(0) = 1]$. See Figure 11.2.

✓   Check that $J_0(t)$ is a solution of Bessel's equation of order 0.

When p is an integer we have to work much harder to get a second solution that is linearly independent of $J_p(t)$. The result is a function $Y_p(t)$ called the *Bessel function of order p of the second kind*. The general formula for $Y_p(t)$ is extremely complicated. We show only the special case $Y_0(t)$, used in the aging spring model:

$$Y_0(t) = \frac{2}{\pi} \left[\left(\gamma + \ln \frac{t}{2} \right) J_0(t) + \sum_{n=0}^{\infty} \frac{(-1)^{n+1} H_n}{(n!)^2} \left(\frac{t}{2} \right)^{2n} \right]$$

☞ Actually γ is an unending decimal (or so most mathematicians believe), and 0.5772 gives the first four digits.

where $H_n = 1 + (1/2) + (1/3) + \cdots + (1/n)$ and γ is *Euler's constant*: $\gamma = \lim_{n \to \infty} (H_n - \ln n) \approx 0.5772$.

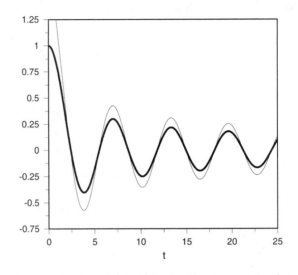

Figure 11.2: The graph of $J_0(t)$ [dark] looks like the graph of the decaying sinusoid $\sqrt{2/\pi t}\cos(t - \pi/4)$ [light].

The general solution of Bessel's equation of integer order p is

$$x(t) = c_1 J_p(t) + c_2 Y_p(t) \tag{10}$$

for arbitrary constants c_1 and c_2. An important thing to note here is that the value of $Y_p(t)$ at $t = 0$ *does* reflect the singularity at $t = 0$; in fact, $Y_p(t) \to -\infty$ as $t \to 0^+$, so that a solution having the form given in (10) is bounded only if $c_2 = 0$.

Bessel functions appear frequently in applications involving cylindrical geometry and have been extensively studied. In fact, except for the functions you studied in calculus, Bessel functions are the most widely used functions in science and engineering.

◆ Transforming Bessel's Equation to the Aging Spring Equation

☞ See "Aging Springs" in Module 11.

Bessel's equation of order zero can be transformed into the aging spring equation $x'' + e^{-at}x = 0$. To do this, we take

$$t = (2/a)\ln(2/as) \tag{11}$$

where the new independent variable s is assumed to be positive. Then we can use the chain rule to find the first two derivatives of the displacement x of the aging spring with respect to s:

$$\frac{dx}{ds} = \frac{dx}{dt}\frac{dt}{ds} = \frac{dx}{dt}\left(-\frac{2}{as}\right)$$

$$\frac{d^2x}{ds^2} = \frac{d}{ds}\left[\frac{dx}{dt}\right]\left(-\frac{2}{as}\right) + \frac{dx}{dt}\frac{d}{ds}\left(-\frac{2}{as}\right)$$

$$= \frac{d^2x}{dt^2}\frac{dt}{ds}\left(-\frac{2}{as}\right) + \frac{dx}{dt}\frac{2}{as^2}$$

$$= \frac{d^2x}{dt^2}\left(-\frac{2}{as}\right)\left(-\frac{2}{as}\right) + \frac{dx}{dt}\frac{2}{as^2}$$

$$= \frac{d^2x}{dt^2}\frac{4}{(as)^2} + \frac{dx}{dt}\frac{2}{as^2}$$

☞ We use w in place of x in the aging spring section of Module 11.

Bessel's equation of order $p = 0$ is given by:

$$s^2\frac{d^2x}{ds^2} + s\frac{dx}{ds} + s^2x = 0$$

and when we substitute in the derivatives we just found, we obtain

$$s^2\left(\frac{d^2x}{dt^2}\frac{4}{(as)^2} + \frac{dx}{dt}\frac{2}{as^2}\right) + s\frac{dx}{dt}\left(-\frac{2}{as}\right) + s^2x = 0$$

Using the fact that

$$s = (2/a)e^{-at/2} \tag{12}$$

(found by solving equation (11) for s) in the last term, when we simplify this monster equation it collapses down to a nice simple one:

$$\frac{d^2x}{dt^2}\frac{4}{a^2} + \frac{4}{a^2}e^{-at}x = 0$$

Finally, if we divide through by $4/a^2$, we get the aging spring equation, $x'' + e^{-at}x = 0$.

The other way around works as well, that is, a change of variables will convert the aging spring equation to Bessel's equation of order zero. That means that solutions of the aging spring equation can be expressed in terms of Bessel functions. This can be accomplished by using $x = c_1 J_0(s) + c_2 Y_0(s)$ as the general solution of Bessel's equation of order 0, and then using formula (12) to replace s. Take another look as Experiments 3 and 4 on Screens 2.5 and 2.6 of Module 11. That will give you a graphical sense about the connection between aging springs and a Bessel's equation.

References

Borrelli, R. L., and Coleman, C. S., *Differential Equations: A Modeling Perspective*, (1998: John Wiley & Sons, Inc.)

Boyce, W. E., and DiPrima, R. C., *Elementary Differential Equations and Boundary Value Problems*, 6th ed., (1997: John Wiley & Sons, Inc.)

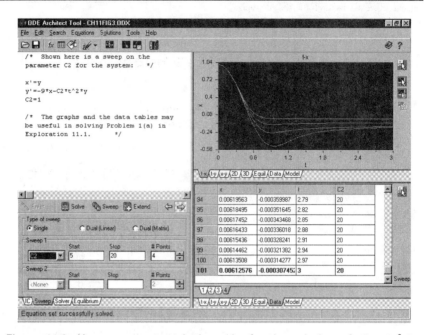

Figure 11.3: Here are some typical graphs for the solution of $x'' + C_2 t^2 x' + 9x = 0$ for various values of C_2. The graphs and the data tables are useful in Problem 1 of Exploration 11.1.

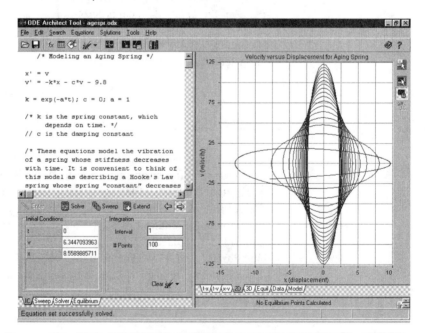

Figure 11.4: Here is a phase-plane portrait for an aging spring ODE, $x'' + e^{-t}x = -9.8$. See "Modeling an Aging Spring" in the library folder "Physical Models" and also Problem 1 in Exploration 11.3.

Answer questions in the space provided, or on
attached sheets with carefully labeled graphs. A
notepad report using the Architect is OK, too.

Name/Date _____

Course/Section _____

Exploration 11.1. Damping a Robot Arm

In each of the following problems it is assumed that the displacement x of a robot arm satisfies an IVP of the form

$$x'' + b(t)x' + 9x = 0, \quad x(0) = 1, \quad x'(0) = 0$$

An optimal damping function $b(t)$ is one for which the solution $x(t)$ reaches 0.005 in minimal time t^* without ever going below zero.

1. Consider damping functions of the form $b(t) = C_k t^k$. For a positive integer k, let C_k^* be the value of C_k that gives the optimal solution, and denote the corresponding minimal time by t_k^*. In Module 11, Screen 1.4 and TTA 3 on Screen 1.7 you found that the optimal solution for $k = 1$ is $x(t) = e^{-9t^2/2}$, with $C_1^* = 9$ and $t_1^* \approx 1.0897$.

   (a) Use ODE Architect to find an approximate optimal solution and values of C_k^* and t_k^* when $k = 2$. [*Suggestion:* Look at Figure 11.3.]

   (b) Repeat with $k = 3$.

   (c) Compare the optimal damping functions for $k = 1, 2, 3$, in the context of the given physical process.

2. For *quadratic damping*, $b(t) = C_2 t^2$, derive a power series solution $x(t) = \sum_{n=0}^{\infty} a_n t^n$. Show that the recurrence formula for the coefficients is

$$a_{n+2} = \frac{-[9a_n + C_2(n-1)a_{n-1}]}{(n+1)(n+2)}, \quad n \geq 1$$

and $a_2 = -9a_0/2$. Recall that $a_0 = x(0)$ and $a_1 = x'(0)$.

3. Let $P_6(t)$ be the Taylor polynomial $\sum_{n=0}^{6} a_n t^n$, where the a_n are given by the recurrence formula in Problem 2.

(a) Write out $P_6(t)$ with C_2 as a parameter; briefly describe how the graph of $P_6(t)$ changes as C_2 increases.

☞ You will need results from Problem 1(a) here.

(b) Graph the apparently optimal solution from Problem 1(a) over the interval $0 \le t \le t_2^*$ and compare it to the graph of $P_6(t)$ with $C_2 = C_2^*$.

4. If the robot arm is totally undamped, its position at time t is $x(t) = \cos 3t$; therefore the arm cannot reach $x = 0$ for all t, $0 \le t \le \pi/6$. In this situation the undamped arm can't remain above $x = 0$. The optimal damping functions $C_k^* t^k$ found in Problem 1 look more like step functions as the degree k increases. Try to improve the time t^* by using a step function for damping.

Assume the robot arm is allowed to fall without damping until just before it reaches $x = 0$, at which time a constant damping force is applied. This situation can be modeled by defining

$$b(t) = \begin{cases} 0 & \text{for } 0 \le t < \frac{\pi}{6} - \varepsilon \\ B_\varepsilon & \text{for } t \ge \frac{\pi}{6} - \varepsilon \end{cases}$$

for $\varepsilon = 0.2, 0.1$, and 0.05. Use ODE Architect to find values of B_ε that give an approximate optimal solution. Include a graph showing your best solution for each ε and give your best value of t^* in each case. What happens to the "optimal" B_ε as $\varepsilon \to 0$?

5. Find a formula for the solution for the situation in Problem 4. The value of ε should be treated as a parameter. Assume that $x(t) = \cos 3t$ for $t < (\pi/6) - \varepsilon$. Then the IVP to be solved is

$$x'' + B_\varepsilon x' + 9x = 0$$
$$x(\pi/6 - \varepsilon) = \cos[3(\tfrac{\pi}{6} - \varepsilon)] = \sin 3\varepsilon$$
$$x'(\pi/6 - \varepsilon) = -3\sin[3(\tfrac{\pi}{6} - \varepsilon)] = -3\cos 3\varepsilon$$

The solution will be of the form $x(t) = c_1 e^{r_1 t} + c_2 e^{r_2 t}$, $r_1 < r_2 < 0$, but the optimal solution requires that $c_2 = 0$. Why? For a fixed ε, find the value of B_ε so that $x(t)$ remains positive and reaches 0.005 in minimum time.

Answer questions in the space provided, or on attached sheets with carefully labeled graphs. A notepad report using the Architect is OK, too.

Name/Date _____

Course/Section _____

Exploration 11.2. Bessel Functions

1. Bessel functions resemble decaying sinusoids. Let's compare the graph of $J_0(t)$ with that of one of these sinusoids.

    (a) On the same set of axes, graph the Bessel function $J_0(t)$ and the function

    $$\sqrt{\frac{2}{\pi t}} \cos\left(t - \frac{\pi}{4}\right)$$

    over the interval $0 \le t \le 10$.

    (b) Now graph these same two functions over the interval $0 \le t \le 50$.

    (c) Describe what you see.

    [*Suggestion:* You can use ODE Architect to plot a good approximation of $J_0(t)$ by solving an IVP involving Bessel's equation in system form:

    $$x' = y, \quad y' = -x - y/t, \qquad x(t_0) = 1, \quad x'(t_0) = 0$$

    with $t_0 = 0.0001$. Actually, $J_0(0) = 1$ and $J_0'(0) = 0$, but $t_0 = 0$ is a singular point of the system so we must move slightly away from zero. You can plot the decaying sinusoid on the same axes as $J_0(t)$ by entering $a = \sqrt{\frac{2}{\pi t}} \cos(t - \frac{\pi}{4})$ in the same equation window as the IVP, selecting a custom 2D plot, and plotting both a and x vs. t.]

2. Repeat Problem 1 for the functions $Y_0(t)$ and $\sqrt{\frac{2}{\pi t}} \sin\left(t - \frac{\pi}{4}\right)$. To graph a good approximation of $Y_0(t)$, solve the system equivalent of Bessel's equation of order zero (from Problem 1) with initial data $t_0 = 0.89357$, $x(t_0) = 0$, $x'(t_0) = 0.87942$. As in Problem 1, we have to avoid the singularity at $t_0 = 0$, especially here because $Y_0(0) = -\infty$. The given initial data are taken from published values of Bessel functions and their derivatives.

Answer questions in the space provided, or on attached sheets with carefully labeled graphs. A notepad report using the Architect is OK, too.

Name/Date _____

Course/Section _____

Exploration 11.3. Aging Spring Models

1. Check out the Library file "Modeling an Aging Spring" in the "Physical Models" folder (see Figure 11.4). The ODE in the file models the motion of a vertically suspended damped and aging spring that is subject to gravity. Carry out the suggested explorations.

2. Show that

$$x(t) = \sqrt{\frac{t+1}{3}} \sin\left(\frac{\sqrt{3}}{2}\ln(t+1)\right) - \sqrt{t+1}\cos\left(\frac{\sqrt{3}}{2}\ln(t+1)\right)$$

    is an analytic solution of the initial value problem

    $$x''(t) + \frac{x(t)}{(t+1)^2} = 0, \quad x(0) = -1, \quad x'(0) = 0$$

    Explain why this IVP provides another model for the motion of an aging spring that is sliding back and forth (without damping) on a support table.

3. Graph the solution $x(t)$ from Problem 2 over the interval $0 \leq x \leq 10$ and compare the graph to the one obtained in Module 11 using ODE Architect.

Answer questions in the space provided, or on attached sheets with carefully labeled graphs. A notepad report using the Architect is OK, too.

Name/Date _____

Course/Section _____

Exploration 11.4. The Incredible Lengthening Pendulum

☞ The ODE for a pendulum of varying length is derived in Chapter 10.

Suppose that we have an undamped pendulum whose length $L = a + bt$ increases linearly over time. Then the ODE that models the motion of this pendulum is

$$(a + bt)\theta''(t) + 2b\theta'(t) + g\theta(t) = 0 \qquad (13)$$

where θ is small enough that $\sin\theta \approx \theta$, the mass of the pendulum bob is 1, and the value of the acceleration due to gravity is $g = 32$.[1]

1. With $a = b = 1$ and initial conditions $\theta(0) = 1$ and $\theta'(0) = 0$, use ODE Architect to solve ODE (13) numerically. What happens to $\theta(t)$ as $t \to +\infty$?

2. Under the same conditions, what happens to the oscillation time of the pendulum as $t \to +\infty$? (The oscillation time is the time between successive maxima of $\theta(t)$.)

[1] See the article "Poe's Pendulum" by Borrelli, Coleman, and Hobson in *Mathematics Magazine*, Vol. 58 (1985) No. 2, pp. 78–83. See also "Child on a Swing" in Module 10.

3. Show that the change of variables

$$s = (2/b)\sqrt{(a+bt)g}, \quad x = \theta\sqrt{a+bt}$$

transforms Bessel's equation of order 1

$$s^2\frac{d^2x}{ds^2} + s\frac{dx}{ds} + (s^2 - 1)x = 0$$

into ODE (13) for the lengthening pendulum. [*Suggestion:* Take a look at the section "Transforming Bessel's Equation to the Aging Spring Equation" in this chapter to help you get started. Use the change of variables given above to express the solution of the IVP in Problem 1 using Bessel functions.]

12 Chaos and Control

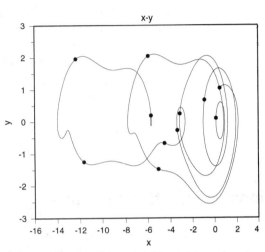

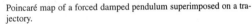

Poincaré map of a forced damped pendulum superimposed on a trajectory.

Overview In this Chapter we'll look at solutions of a forced damped pendulum ODE. In the linear approximation of small oscillations, this ODE becomes the standard constant-coefficient ODE $x'' + cx' + kx = F(t)$, which can be solved explicitly in all cases. Without the linear approximation, the pendulum ODE contains the term $k \sin x$ instead of kx. Now the study becomes much more complicated. We'll focus on the special case of the nonlinear pendulum ODE

$$x'' + cx' + \sin x = A \cos t \tag{1}$$

but our results leave a world of further things to be discovered. We'll show that appropriate initial conditions will send the pendulum on any desired sequence of gyrations, and hint at how to control the chaos by finding such an initial condition.

Key words Forced damped pendulum; sensitivity to initial conditions; chaos; control; Poincaré sections; discrete dynamical systems; Lakes of Wada; control

See also Chapter 10 for background on the pendulum. Chapter 13 for more on discrete dynamical systems and other instances of chaos and sensitivity to initial conditions.

◆ Introduction

How might chaos and control possibly be related? These concepts appear at first to be opposites, but in fact they are two faces of the same coin!

A good way to start discussing this apparent paradox is to think about learning to ski. The beginning skier tries to be as stable as possible, with feet firmly planted far enough apart to give confidence that she or he will not topple over. If you try to ski in such a position, you cannot turn, and the only way to stop, short of running into a tree, is to fall down. Learning to ski is largely a matter of giving up on "stability," bringing your feet together so as to acquire controllability! You need to allow chaos in order to gain control.

Another example of the relation between chaos and control is the early aircraft available at the beginning of World War I, carefully designed for greatest stability. The result was that their course was highly predictable, an easy target for antiaircraft fire. Very soon the airplane manufacturers started to build in enough instability to allow maneuverability!

◆ Solutions as Functions of Time

The methods of analysis we will give can be used for many other differential equations, such as Duffing's equation

$$x'' + cx' + x - x^3 = A \cos \omega t, \tag{2}$$

or the differential equation

$$x'' + cx' + x - x^2 = A \cos \omega t, \tag{3}$$

which arises when studying the stability of ships. The explorations at the end suggest some strategies for these problems.

Let's begin to study ODE (1) with $c = 0.1$:

$$x'' + .1x' + \sin x = A \cos t \tag{4}$$

Let's compute some solutions, starting at $t = 0$ with $A = 1$ and various values of $x(0)$ and $x'(0)$, and observe the motion out to $t = 100$, or perhaps longer (see Figure 12.1). We see that most solutions eventually settle down to an oscillation with period 2π (the same period as the driving force). This xt-plot actually shows oscillations which differ by multiples of 2π.

This settling down of behaviors at various levels is definitely a feature of the parameter values chosen: for the amplitude $A = 2.5$ in ODE (4), for instance, there does not appear to be any steady-state oscillation at all.

Looking at such pictures is quite frustrating: it is very hard to see the pattern for which initial conditions settle down to which stable oscillations, and which will not settle down at all.

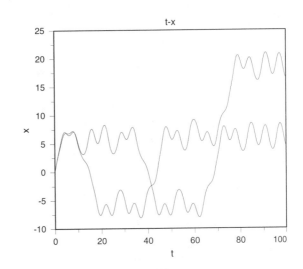

Figure 12.1: Solution curves of ODE (4) with $x(0) = 0$, $x'(0) = 2, 2.1$.

◆ Poincaré Sections

Poincaré found a way to understand and visualize the behavior of our differential equation: he sampled solutions of ODE (4) at multiples of the period 2π of the driving function:

$$0, 2\pi, 4\pi, \ldots, 2k\pi, \ldots$$

This is much like taking pictures with a strobe light.

An equivalent way of saying this is to say that we will iterate[1] the mapping $P : \mathbb{R}^2 \to \mathbb{R}^2$ which takes a point (a, b) in $\mathbb{R}^2$, computes the solution $x(t)$ with $x(0) = a$, $x'(0) = b$, and sets

☞ Note that the clock starts at $t_0 = 0$ when generating Poincaré plots.

$$P(a, b) = (x(2\pi), x'(2\pi)) \tag{5}$$

This mapping P is called a *Poincaré mapping*. If you apply the operator P to (a, b) k times in succession, the result is $P^k(a, b)$ and we see that

$$P^k(a, b) = (x(2k\pi), x'(2k\pi))$$

☞ When the xx'-plane is used to chart the evolution of the points $P^k(a, b), k = 1, 2, \ldots$, it is called the Poincaré plane.

In a sense, the Poincaré section is simply a crutch: every statement about Poincaré sections corresponds to a statement about the original ODE, and vice versa. But this crutch is invaluable since the orbits of a nonautonomous ODE such as ODE (4) often intersect each other and themselves in a hopelessly tangled way.

---

[1]Chapter 13 discusses iterating maps $f : \mathbb{R} \to \mathbb{R}$; there you will find that the map $f(x) = \lambda x(1 - x)$ is filled with surprises. Before trying to understand the iteration of P, which is quite complicated indeed, the reader should experiment with several easier examples, like linear maps of $\mathbb{R}^2 \to \mathbb{R}^2$.

◆ Periodic Points

A good way to start investigating the Poincaré mapping P (or for that matter, the iteration of any map) is to ask: what periodic points does it have? Setting $x' = y$, a *periodic point* is a point (x, y) in $\mathbb{R}^2$ such that for some integer k we have $P^k(x, y) = (x, y)$. *Fixed* points are periodic points with $k = 1$, and are particularly important.

Periodic points of period k for P are associated with periodic solutions of ODE (4) of period $2k\pi$. In particular, if $x(t)$ is a solution which is periodic of period 2π, then

$$(x(0), x'(0)) = (x(2\pi), x'(2\pi))$$

is a fixed point of P. If you observe this solution with a strobe which flashes every 2π, you will always see the solution in the same place.

◆ The Unforced Pendulum

If there is no forcing term in ODE (4), then we have an autonomous ODE like those treated in Chapter 10.

Example: The ODE

$$x'' + x' + 10 \sin x = 0$$

models a damped pendulum without forcing. A phase plane portrait is shown in Figure 12.2. Note that the equilibrium points (of the equivalent system) at $x = 2n\pi$, $x' = 0$ are spiral sinks, but the equilibrium points at $x = (2n + 1)\pi$, $x' = 0$ are saddles. Note also that the phase plane is divided into slanting regions, each of which has the property that its points are attracted to the equilibrium point inside the region. These regions are called *basins of attraction*. If a forcing term is supplied, these basins become all tangled up (Figure 12.4 on page 227).

There is a Poincaré mapping P for the unforced damped pendulum, which is fairly easy to understand, and which you should become familiar with before tackling the forced pendulum. In this case, two solutions of ODE (4) with $A = 0$ stand out: the equilibria $x(t) = 0$ and $x(t) = \pi$ for all t. Certainly if the pendulum is at one of these equilibria and you illuminate it with a strobe which flashes every T seconds, where T is a positive number, you will always see the pendulum in the same place. Thus these points are fixed points of the corresponding Poincaré mapping P. In the xx'-plane, the same thing happens at the other equilibrium points, that is, at the points $\dots, (-2\pi, 0), (0, 0)(2\pi, 0), \dots$ for the "downward" stable equilibria, and at the points $\dots, (-3\pi, 0), (-\pi, 0), (\pi, 0), \dots$ for the unstable equilibria.

The analysis in Module 10 using an integral of motion should convince you that for the unforced damped pendulum, these are the only periodic points: if the pendulum is not at an equilibrium, the value of the integral decreases with time, and the system cannot return to where it was.

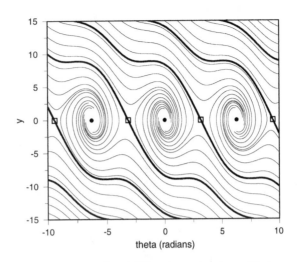

Figure 12.2: Basins of attraction of the downward equilibrium positions of the unforced damped pendulum are bounded by separatrices.

If you start the pendulum with both $x(0)$ and $x'(0)$ small, the damping will simply kill off the motion, and the pendulum will be attracted to the downward equilibrium. The point $(0, 0)$ in state space is called a *sink*.

The behavior is more interesting near an unstable equilibrium. Imagine imparting an initial velocity to the bob by kicking it. For a small kick, it will swing back. Now kick it a little harder: it will rise higher, and still swing back. Kick it harder still, and it will make it over the top, and hit you in the back if you aren't careful. Dividing the kicks which don't make it over from those that do is a very special kick, where the pendulum rises forever, more and more slowly, tending to the unstable equilibrium. Thus there are initial conditions which generate solutions that tend to the unstable equilibrium; in the Poincaré plane these solutions form two curves which meet end to end at the fixed point corresponding to the unstable equilibrium. Together they form the *stable separatrix* of the fixed point. There are also curves of initial conditions which come from the unstable equilibrium; together they form the *unstable separatrix* of the equilibrium. See Figure 12.3.

As stated earlier, a good first thing to do when iterating a map is to search for the periodic points; a good second thing to do is to find the periodic points which correspond to unstable equilibria (*saddles*, in the case of the pendulum) and find their separatrices.

For the unforced damped pendulum, the equilibria of the differential equation and the fixed points of any Poincaré map coincide; so, too, do the separatrices of the unstable equilibria (in the phase plane) and the separatrices of the corresponding saddle fixed points in the Poincaré plane. These separatrices separate the trajectories which approach a given sink from the trajectories that approach a different sink.

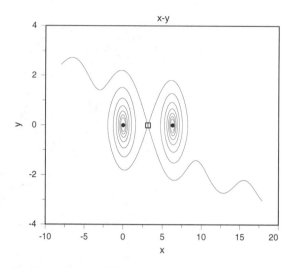

Figure 12.3: Stable and unstable separatrices at a saddle for an unforced damped pendulum. Which are the stable separatrices?

✓   "Check" your understanding by reproducing the plot in Figure 12.3.

◆ The Damped Forced Pendulum

We described above the Poincaré plane for the unforced pendulum. The same description holds for the forced pendulum. A figure showing a Poincaré map for a forced pendulum appears as the chapter cover figure. Thus, in the Poincaré plane, we expect to see a collection of fixed points corresponding to the oscillations to which the pendulum "settles down", and each has a basin: the set of initial conditions which will settle down to it. The basins appear to be extraordinarily tangled and complicated, and they are. The reader should put up the picture of the basins (Screen 2.6 in Module 12), and practice superimposing iterations on the figure, checking that if you start in the blue basin, the entire orbit remains in the blue basin, perhaps taking a complicated path to get near the sink, but making it in the end.

◆ Tangled Basins, the Wada Property

In the tangled basins Screen 3.3 of Module 12, each basin appears to be made of a central piece, and four canals which go off and meander around the plane. The meandering appears to be completely random and chaotic, and the only thing the authors really know about the shapes of the basins of our undamped pendulum is the following fact: The basins have the *Wada property*: every

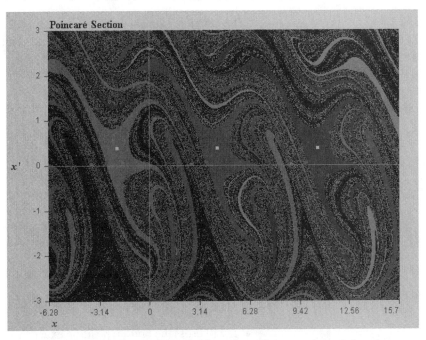

Figure 12.4: Tangled basins for a forced damped pendulum.

point of the boundary[2] of any basin is in the boundary of all the others. Thus if you start at a boundary point of any basin, and perturb the initial condition an arbitrarily small amount, you can land in any of the infinitely many basins.

A careful look at Figure 12.4 should convince you that this stands a good chance of being true: Each region of a canal boundary point includes pieces on many curves. It isn't clear, of course, that there are canals of *all* the basins between any two canals.

It is one thing to think that the Wada property is likely true, and quite another to prove it. It isn't clear how you would prove anything whatsoever about the basins: they do not appear to be amenable to precise study.

To get a grip on these basins, the first step is to understand why they appear to be bounded by smooth curves, and to figure out what these smooth curves are. For each sink (solid white squares in Figure 12.4), there are in fact four periodic points, each of period two, which are saddles, and such that for each saddle one of the two unstable separatrices is entirely contained in the corresponding basin.

---

[2]The boundary ∂U of an open set $U \subset \mathbb{R}^2$ is a point $\mathbf{x} \in \mathbb{R}^2$ which is not in U, but such that there exists a sequence of points $\mathbf{x}_n \in U$ which converges to $\mathbf{x}$. Later we will encounter the notion of *accessible boundary*: the points $\mathbf{x} \in \partial U$ such that there exists a parametrized curve $\gamma : (0, 1] \to U$ such that $\lim_{t \to 0} \gamma(t) = \mathbf{x}$. For simple open sets, the boundary and the accessible boundary coincide, but not for our basins.

The next step is to show that the *accessible boundary* of the basin is made up of the stable separatrices of these saddles. This uses the technique of *basin cells*, as pioneered by Kennedy, Nusse and Yorke[3]. To see a fleshed out sketch, see the C·ODE·E article referenced at the end of this chapter.

◆ Gaining Control

The statement about the basins having the Wada property is, in some sense, a negative statement, saying that there is maximum possible disorder. Is there some positive statement one can make about the forced pendulum (for these parameter values)? It turns out that there is. The precise statement is as follows.

During one period of the forcing term, say during

$$t \text{ in the interval } I_k = [2k\pi, 2(k+1)\pi]$$

the pendulum will do one of the following four things:

- It will cross the bottom position exactly once moving clockwise (count this possibility as -1);
- It will cross the bottom position exactly once moving counterclockwise (count this possibility as $+1$);
- It will not cross the bottom position at all (count this possibility as 0);
- It will do something else (possibility NA).

Note that most solutions appear to be attracted to sinks, and that the stable oscillation corresponding to a sink crosses the bottom position twice during each I_k, and hence these oscillations (and most oscillations after they have settled down) belong to the NA category.

The essential control statement we can make about the pendulum is the following:

For any biinfinite sequence $\dots, \varepsilon_{-1}, \varepsilon_0, \varepsilon_1, \dots$ of symbols ε_i selected from the set $\{-1, 0, 1\}$, there exists $x(0), x'(0)$ such that the solution with this initial condition will do ε_k during the time interval I_k.

The chaos game in Module 12 suggests why this might be true; the techniques involved in the proof were originally developed by Smale[4].

---

[3] Judy Kennedy, Helena Nusse, and James Yorke are mathematicians at the Universities of Delaware, Utrecht, and Maryland, repectively.

[4] Stephen Smale is a contemporary mathematician who was awarded a Fields medal (the mathematical equivalent of a Nobel prize) in the early 1960's. See Devaney in the references at the end of this chapter.

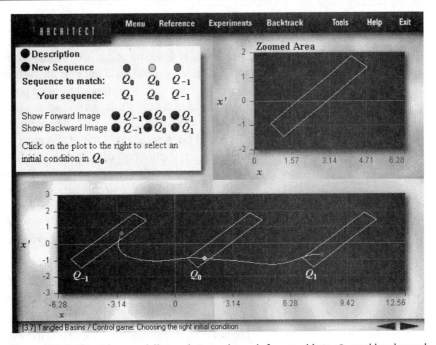

Figure 12.5: Start in quadrilateral Q_0 and reach forward into Q_1 and backward into Q_{-1}.

We start by drawing quadrilaterals Q_k around the kth saddle, long in the unstable direction and short in the stable direction, such that they cross a good part of the tangle. We can now translate our symbols ε_i, which refer to the differential equation, into the Poincaré mapping language.

If at time $t = 2k\pi$ the pendulum is in Q_k and at time $2(k+1)\pi$ it is in $Q_{k+\varepsilon_k}$, then during I_k the pendulum does ε_k. So it is the same thing to require that a trajectory of the pendulum realize a particular symbol sequence, and to require that an orbit of the Poincaré map visit a particular sequence of quadrangles, just so long as successive quadrangles be neighbors or identical.

Draw the forward image of that quadrilateral, and observe that it grows much longer in the unstable direction and shrinks in the stable direction; we will refer to $P(Q_k)$ as the kth *snake*, S_k. The entire proof comes down to understanding how S_k intersects Q_{k-1}, Q_k and Q_{k+1}.

The thing to be checked is that S_k intersects all three in subquadrangles going from top to bottom, and that the top and bottom of Q_k map to parts of the boundary of S_k which are outside $Q_{k-1} \cup Q_k \cup Q_{k+1}$. See Figure 12.5 for an example of a winning strategy for three adjacent quadrilaterals.

Once you have convinced yourself that this is true, you will see that every symbolic sequence describing a history of the pendulum is realized by an intersection of thinner and thinner nested subquadrangles.

A similar argument shows that a symbolic sequence describing a future of the pendulum corresponds to a sequence of thinner and thinner subquadrangles going from left to right. The details are in the C·ODE·E paper by

J.H. Hubbard in the references.

We have shown how to gain control of the motions of the driven pendulum. In particular, if we want the pendulum "robot" to execute a prescribed set of rotations, all we have to do is put it in the right initial state and switch on the driving force. Although everything has been phrased in terms of a pendulum, the approach extends to almost any kind of chaotic motion. Engineers, scientists, and mathematicians are now designing prototypes of chaotic control systems based on these ideas. One of the most intriguing applications uses a chaotic sequence to encode a digital message. The sequence is "added" to the message, and the intended recipient then subtracts the chaos to read the message. Chaos is often an undesirable aspect of physical motions. Devices have recently been built that force a chaotic system to stay focussed on a desired response. All of this is new, so we can't say just how the applications will evolve. See the books by Kapitaniak, Nayfeh, and Ott for more on chaotic controls and controlling chaos.

References

Borrelli, R. and Coleman, C., "Computers, Lies, and the Fishing Season" in *The College Mathematics Journal*, November 1994, pp. 403–404

Devaney, R. L., *An Introduction to Chaotic Dynamical Systems*, (1986: Benjamin/Cummings), Section 2.3 "The Horseshoe Map"

Guckenheimer, J. and Holmes, P., *Nonlinear Oscillations, Dynamical Systems, and Bifurcations of Vector Fields* (1983: Springer-Verlag). The original classic in this field.

Hastings, S.P. and MacLeod, J.B., "Chaotic Motion of a Pendulum with Oscillatory Forcing" in *The American Mathematical Monthly*, June-July 1993

Hubbard, J.H., "What It Means to Understand a Differential Equation" in *The College Mathematics Journal*, November 1994, pp. 372–384

Hubbard, J.H., "The Forced Damped Pendulum: Chaos, Complexity, and Control" in *C·ODE·E Newsletter*, Spring 95; (soon to be published in *The American Mathematical Monthly*)

Kapitaniak, T., *Controlling Chaos* (1996, Academic Press)

Nayfeh, A.H. and Balachandran, B., *Applied Nonlinear Dynamics* (1995, John Wiley & Sons, Inc.)

Ott, E., Sauer, T., Yorke, J.A., *Coping with Chaos* (1994, John Wiley & Sons, Inc.)

Sharp, J. "A Problem in Ship Stability" Group project for interdisciplinary course in nonlinear dynamics and chaos, 1996-1997. Write to John Sharp, Department of Physics, Rose-Hulman University, Terre Haute, IN.

Strogatz, S., *Nonlinear Dynamics and Chaos*, (1994: Addison-Wesley). Nice treatment of many problems, including chaos induced in constant torque motion.

Answer questions in the space provided, or on
attached sheets with carefully labeled graphs. A
notepad report using the Architect is OK, too.

Name/Date _____

Course/Section _____

Exploration 12.1.

In each problem describe what you see and explain what the figures tell you
about the behavior of the pendulum.

1. Choose a value for $c \neq 0.1$, take $A = 1$ in ODE (1), and produce graphs like
those in the chapter cover figure and Figure 12.1.

2. Choose a value for $A \neq 1$ and $c = 0.1$ in ODE (1) and produce graphs like
those in the chapter cover figure and Figure 12.1.

3. Choose a value for $\omega \neq 1$ in the ODE

$$x'' + 0.1x' + \sin x = \cos \omega t$$

and produce graphs like those in the chapter cover figure and Figure 12.1.

4. Repeat Problems 1 and 2, but for the Duffing ODE,

$$x'' + cx' + x - x^3 = A\cos t$$

5. Repeat Problems 1 and 2, but for the ODE with a quadratic nonlinearity,

$$x'' + cx' + x - x^2 = A\cos t$$

13 Discrete Dynamical Systems

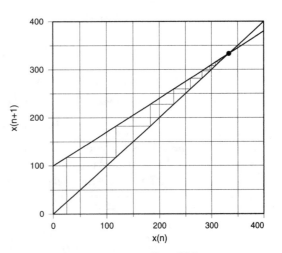

Supply and demand converge to a stable equilibrium.

Overview   Processes such as population dynamics that evolve in discrete time steps are best modeled using discrete dynamical systems. These take the form $x_{n+1} = f(x_n)$, where the variable x_n is the state of the system at "time" n and x_{n+1} is the state of the system at time $n + 1$. Discrete dynamical systems are widely used in ecology, economics, physics and many other disciplines. In this section we present the basic techniques and phenomena associated with discrete dynamical systems.

Key words   Iteration; fixed point; periodic point; cobweb and stairstep diagrams; stability; sinks; sources; bifurcation diagrams; logisitic maps; chaos; sensitive dependence on initial conditions; Julia sets; Mandelbrot sets

See also   Chapter 6 for more on sinks and sources in differential equations; Chapter 12 for Poincaré sections.

Discrete dynamical systems arise in a large variety of applications. For example, the population of a species that reproduces on an annual basis is best modeled using discrete systems. Discrete systems also play an important role in understanding many *continuous* dynamical systems. For example, points calculated by a numerical ODE solver form a discrete dynamical system that approximates the solution of an initial value problem for an ODE. The Poincaré section described in Chapter 12 is another example of a discrete dynamical system that gives information about a system of ODEs.

A *discrete* dynamical system is defined by the *iteration* of a function f, and takes the form

$$x_{n+1} = f(x_n), \quad n \geq 0, \quad x_0 \text{ given} \tag{1}$$

Here are another two examples. In population dynamics, some populations are modeled using a *proportional growth* model

☞ The function $f(x) = \lambda x$ is denoted L_λ, and so $L_\lambda(x) = \lambda x$.

$$x_{n+1} = L_\lambda(x_n) = \lambda x_n, \quad n \geq 0, \quad x_0 \text{ given} \tag{2}$$

where x_n is the population density at generation n and λ is a positive number that measures population growth from generation to generation. Another common model is the *logistic growth* model:

☞ The function $\lambda x(1-x)$ is denoted by $g_\lambda(x)$.

$$x_{n+1} = g_\lambda(x_n) = \lambda x_n(1 - x_n), \quad n \geq 0, \quad x_0 \text{ given}$$

Let's return to the general discrete system (1). Starting with an initial condition x_0, we can generate a *sequence* using this rule for iteration: Given x_0, we get $x_1 = f(x_0)$ by evaluating the function f at x_0. We then compute $x_2 = f(x_1)$, $x_3 = f(x_2)$, and so on, generating a sequence of points $\{x_n\}$.

☞ The superscript ° reminds us that this is just the composition of f with itself; f is *not* being raised to a power.

Each x_n is the *n-fold composition* of f at x_0 since

$$x_2 = f(f(x_0)) = f^{\circ 2}(x_0)$$
$$x_3 = f(f(f(x_0))) = f^{\circ 3}(x_0)$$
$$\vdots$$
$$x_n = f^{\circ n}(x_0)$$

(Some authors omit the superscript °.)

The infinite sequence of iterates $O(x_0) = \{x_n\}_{n=0}^{\infty}$ is called the *orbit of x_0 under f*, and the function f is often referred to as a *map*. For example, if we take $\lambda = 1/2$ and the initial condition $x_0 = 1$ in the proportional growth model (2), we get the orbit for the map L:

$$x_0 = 1, \quad x_1 = 1/2, \quad x_2 = 1/4, \ldots$$

Refer to Screen 1.2 of Module 13 for four representations of the orbit of an iteration: as a *sequence* $\{x_0, x_1 = f(x_0), x_2 = f(x_1), \ldots\}$; a *numerical list* whose columns are labeled n, x_n, $f(x_n)$; a *time series* where x_n is plotted against "time" n; and a *stairstep/cobweb diagram* for graphical analysis.

The chapter cover figure shows a stairstep diagram for the model $x_{n+1} = 0.7x_n + 100$. Figures 13.1 and 13.2 show cobweb diagrams for the logistic model $x_{n+1} = \lambda x_n (1 - x_n)$, with $\lambda = 3.51$ and 3.9, respectively. In all of these figures, the diagonal line $x_{n+1} = x_n$ is also plotted. The stairstep and cobweb diagrams are constructed by selecting a value for x_0 on the horizontal axis, moving up to the graph of the iterated function to obtain x_1, horizontally over to the diagonal then up (or down) to the graph of the function to obtain x_2, and so on. These diagrams are used to guide the eye in moving from x_n to x_{n+1}.

◆ Equilibrium States

As with autonomous ODEs, it is useful to determine the equilibrium states for a discrete dynamical system. First we need some definitions:

☞ A fixed point of a discrete dynamical system is the analogue of an equilibrium point for a system of ODEs.

- A point x^* is a *fixed point* of f if $f(x^*) = x^*$. A fixed point is easy to spot in a stairstep or cobweb diagram even before the steps and webs are plotted: the fixed points of f are where the graph of f intersects the diagonal.

- A point x^* is a *periodic point of period n* of f if $f^{\circ n}(x^*) = x^*$ and $f^{\circ k}(x^*) \neq x^*$ for $k < n$. A fixed point is a periodic point of period 1.

Both the proportional and logistic growth models have the fixed point $x = 0$. For certain values of λ the logistic model has periodic points; Figure 13.1 suggests that the model has a period-4 orbit if $\lambda = 3.51$.

✓ "Check" your understanding by showing that the logistic model has a second fixed point $x^* = (\lambda - 1)/\lambda$. Does the proportional growth model for $\lambda > 0$ have any periodic points that are not fixed?

☞ This use of the words "stable" and "unstable" for points and orbits of a discrete system differs from the way the words are used for equilibrium points of an ODE. For example, a saddle point of an ODE is unstable, but a saddle point of a discrete system is neither stable or unstable.

A fixed point x^* of f is said to be *stable* (or a *sink*, or an *attractor*) if every point p in some neighborhood of x^* approaches x^* under iteration by f, that is, if $f^{\circ n}(p) \to x^*$ as $n \to +\infty$. The set of *all* points such that $f^{\circ n}(p) \to x^*$ as $n \to +\infty$ is the *basin of attraction* of p. A fixed point x^* is *unstable* (or a *source* or *repeller*) if every point in some neighborhood of x^* moves out of the neighborhood under iteration by f. If x^* is a period-n point of f, then the orbit of x^* is said to be *stable* if x^* is stable as a fixed point of the map $f^{\circ n}$. The orbit is *unstable* if x^* is unstable as a fixed point of $f^{\circ n}$. Stability is determined by the *derivative* of the map f, as the following tests show:

- A fixed point x^* is stable if $|f'(x^*)| < 1$, and unstable if $|f'(x^*)| > 1$.
- The orbit of a periodic point x^* of period n is stable if $|(f^{\circ n})'(x^*)| < 1$, and unstable if $|(f^{\circ n})'(x^*)| > 1$.

Stable periodic orbits are *attracting* because nearby orbits approach them, while unstable periodic orbits are *repelling* because nearby orbits move away from them.

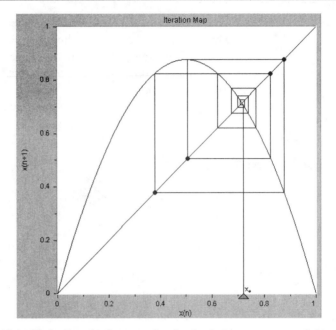

Figure 13.1: The cobweb diagram for the logistic map, $x_{n+1} = 3.51x_n(1 - x_n)$, **suggests that iterates of** $x_0 = 0.72$ **approach a stable orbit of period 4.**

✓ Is the fixed point in the chapter cover figure stable? Is the period-4 orbit in Figure 13.1 stable? How about the two fixed points in that figure? [*Suggestion:* Use the Discrete Tool as an aid in answering these questions.]

◆ Linear versus Nonlinear Dynamics

The solutions of linear and of nonlinear ODEs are compared and contrasted in Chapter/Modules 6 and 7. Now we will do the same comparison for linear and nonlinear maps of the real line into itself.

☞ Refer to the first submodule of Module 13 for examples.

Let's look at the iteration of linear functions such as the proportional growth model $x_{n+1} = L_\lambda(x_n)$, which has a fixed point at $x^* = 0$. This fixed point is stable if $|\lambda| < 1$, so the orbit of every initial population tends to 0 as $n \to \infty$. If $\lambda = 1$, then $x_{n+1} = x_n$, and hence every point is a fixed point. The fixed point at $x^* = 0$ is unstable if $|\lambda| > 1$, and all initial populations tend to ∞ as $n \to \infty$. If $\lambda = -1$ then $x^* = 0$ is the only fixed point and every other point is of period 2 since $x_{n+1} = -x_n$.

The iteration of any linear function $f(x) = ax + b$ (with slope $a \neq 1$) behaves much like the proportional growth model. Fixed points are found by solving $ax^* + b = x^*$, and their stability is governed by the magnitude of a.

The iteration of nonlinear functions can be much more complex than that of linear functions. In particular, nonlinear functions can exhibit chaotic behavior, as well as periodic behavior. To illustrate the types of behavior typical

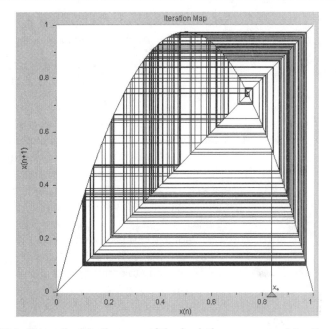

Figure 13.2: The cobweb diagram of the logistic map $x_{n+1} = 3.9x_n(1 - x_n)$ suggests that iterates of $x_0 = 0.8361$ either approach an attracting periodic orbit of very high period, or else wander chaotically.

to nonlinear functions we consider, in the second submodule of Module 13, the one-parameter family of *logistic* functions

$$g_\lambda(x) = \lambda x(1 - x)$$

Figure 13.2 shows how complex an orbit of a logistic map may be for certain values of λ.

◆ Stability of a Discrete Dynamical System

Now we turn our attention to the stability of an entire dynamical *system* rather than just that of a fixed point. One of the most important ideas of dynamical systems (discrete or continuous) is that of *hyperbolicity*. Hyperbolic points are stable to small changes in the parameters of a dynamical system. This does not mean that a perturbation (a small change) of the function leaves the fixed or periodic point unchanged. It simply means that the perturbed function will also have a fixed point or periodic point "nearby," and that this point has the stability properties of the corresponding point of the unperturbed function. For example, at $\lambda = 2$ the function $g_2(x)$ has an attracting fixed point $x^* = 0.5$. For values of λ near 2, the function $g_\lambda(x)$ also has an attracting fixed point $x^* = (\lambda - 1)/\lambda$. For example, if $\lambda = 2.1$ then the attracting fixed point is $x^* = 0.524$. Even though the fixed point moved a little as λ increased, the fixed point still exists and it is still attracting. The following theorem provides a way of determining whether fixed points and periodic orbits are hyperbolic.

THEOREM 13.1   Given a discrete dynamical system $x_{n+1} = f_\lambda(x_n)$, a fixed point x^* of $f_\lambda(x)$ is hyperbolic if $|f_\lambda'(x^*)| \neq 1$. Similarly, a periodic point x^* of period n (and its orbit) is hyperbolic if $|(f_\lambda^{\circ n})'(x^*)| \neq 1$.

Because the number and type of periodic points do not change at parameter values where $f_\lambda(x)$ has hyperbolic points, we say that the qualitative structure of the dynamical system remains unchanged. On the other hand, this theorem also implies that changes in the qualitative structure of a family of discrete dynamical systems can occur only when a fixed or periodic point is *not* hyperbolic. We see this in the proportional growth model $x_{n+1} = \lambda x_n$ when $\lambda = 1$ and $\lambda = -1$. For $\lambda = 1 - \varepsilon$ [and hence $L_{1-\varepsilon}'(1) = \lambda = 1 - \varepsilon$] the fixed point $x = 0$ is attracting. But for $\lambda = 1 + \varepsilon$ the fixed point is repelling. Thus, as λ passes through the value 1, the stability of the fixed point changes from attracting to repelling and the qualitative structure of the dynamical system changes.

◆ Bifurcations

A change in the qualitative structure of a discrete dynamical system, such as a change in the stability of a fixed point, is known as a *bifurcation*. Two other types of bifurcations can also occur when f_λ is nonlinear.

The first, known as a *saddle-node* bifurcation, occurs when x^* is a periodic point of period n and $(f_\lambda^{\circ n})'(x^*) = 1$. In a saddle-node bifurcation, the periodic point x^* splits into a pair of periodic points, both of period n, one of which is attracting and the other repelling. A saddle-node bifurcation occurs in the logistic growth family $g_\lambda(x)$ when $\lambda = 1$. At this value the fixed point $x^* = 0$ (which is attracting for $\lambda < 1$) splits into a pair of fixed points, $x^* = 0$ (repelling for $\lambda > 1$), and $x^* = (\lambda - 1)/\lambda$ (attracting for $\lambda > 1$). This type of bifurcation is sometimes called an *exchange of stability* bifurcation.

The second important type of bifurcation is called *period-doubling* and occurs when x^* is a periodic point of period-n and $(f_\lambda^{\circ n})'(x^*) = -1$. In this bifurcation the attracting period n point becomes repelling and an attracting period-$2n$ orbit is spawned. (Note that the stability can be reversed.) This occurs in the logistic family $g_\lambda(x)$ when $\lambda = 3$. At this parameter value, the attracting fixed point $x^* = (\lambda - 1)/\lambda$ becomes repelling and a stable period-2 orbit emerges with one point on each side of $x^* = (\lambda - 1)/\lambda$. Since the logistic equations model population growth, this says that the population converges to an equilibrium for growth rate constants λ less than 3. However, for values of λ greater than 3, the population oscillates through a sequence of values.

The bifurcations that occur in a one parameter family of discrete dynamical systems can be summarized in a *bifurcation diagram*. For each value of the parameter (on the horizontal axis) the diagram shows the long-term behavior under iteration of a "typical" initial point. For example, if you see

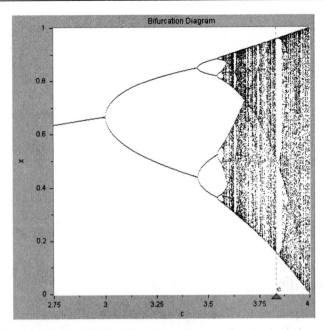

Figure 13.3: Part of the bifurcation diagram for the logistic map.

a single point in the diagram above a particular parameter value, that point corresponds to an attracting fixed point. The spot in the diagram where you see an arc of attracting fixed points split into two arcs corresponds to a bifurcation from an attracting fixed point for an attracting orbit of period 2 (i.e., period doubling). If the diagram shows a multitude of points above a given parameter value, then either you are seeing an attracting periodic orbit of a very high period, or else you are seeing chaotic wandering. It should be noted that when constructing the bifurcation diagram for each parameter value and initial point, the first 50 or so iterates are omitted so that only the long-term behavior is visible in the diagram. See Figure 13.3 and Screen 2.4 in Module 13 for the bifurcation diagram of the logistic map.

The stable arcs in these diagrams are usually straightforward to generate numerically. We constructed a bifurcation diagram on an interval $[\lambda_{min}, \lambda_{max}]$ for the logistic population model $x_{n+1} = g_\lambda(x_n)$ using the following procedure.

1. Fix λ_{min}, λ_{max}, λ_{inc}, n_{min}, n_{max}. Here λ_{inc} is the step size between successive values of λ while n_{min} and n_{max} are bounds on the number of iterates used to construct the diagram; they control the accuracy of the diagram. Typical values are $n_{min} = 50$ and $n_{max} = 150$.

2. Let $\lambda = \lambda_{min}$.

3. Taking $x_0 = 0.5$ for example, compute the first n_{min} iterates of g_λ without plotting anything. This eliminates transient behavior.

4. For $n_{\min} \leq n \leq n_{\max}$, plot the points $(\lambda, g_\lambda^{\circ n}(0.5))$. If the orbit of 0.5 converges to a periodic orbit, only points near this orbit are plotted. If the orbit of 0.5 isn't periodic, then the points above λ seem to be almost randomly distributed.

5. Let $\lambda = \lambda + \lambda_{\mathrm{inc}}$.

6. If $\lambda < \lambda_{\max}$, go back to Step 3 and repeat the process.

✓   Go to the one-dimensional tab of the Discrete Tool. Use the default values, but set the value of c at 1 (c in the tool plays the role of λ in Chapter/Module 13). Click on the bifurcation diagram. Keep your finger on the up-arrow for c and describe what is happening. Any attracting periodic orbits? For what values of c do these orbits occur? What are the periods?

◆ Periodic and Chaotic Dynamics

One of the most celebrated theorems of discrete dynamical systems is often paraphrased "Period 3 Implies Chaos." This theorem, originally proven by Šarkovskii and independently discovered by Li and Yorke[1], is a remarkable result in that it requires little information about the dynamical system and yet it returns a treasure trove of information.

THEOREM 13.2   If f is a continuous function on the real line and if there exists a point of period 3, then there exist points of every period.

For the logistic population model there exists an attracting period-3 orbit at $\lambda = \sqrt{8} + 1 \approx 3.83$, and most initial conditions in the unit interval converge to this orbit (see Figure 13.4). In terms of our model, most populations tend to oscillate between the three different values of the period-3 orbit. Theorem 13.2 states that even more is going on at $\lambda = \sqrt{8} + 1$ than meets the eye. If we pick any positive integer n, there exists a point p such that n is the smallest positive integer allowing $g_\lambda^{\circ n}(p) = p$. Thus, for example, there exists a point that returns to itself in 963 iterates. The reason we don't "see" this periodic orbit (or, indeed, any periodic orbit, except that of period 3) is that it is unstable, so no iterate can approach it. But orbits of every period are indeed present if $\lambda = \sqrt{8} + 1$.

---

[1] James Yorke and T.Y. Li are contemporary mathematicians who published their result in 1975 (see References). They were the first to apply the word "chaos" to the strange behavior of the iterates of functions such as g_λ. A.N. Šarkovskii published a stronger result in 1964, in Russian, in the *Ukrainian Mathematical Journal*, but it remained unknown in the West until after the paper by Yorke and Li had appeared.

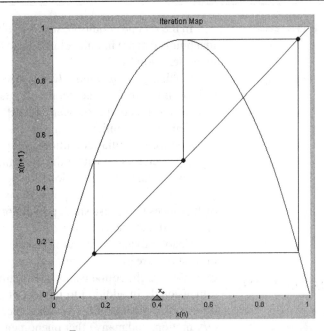

Figure 13.4: At $\lambda = \sqrt{8} + 1 \approx 3.83$ the logistic map $g_\lambda = \lambda x(1 - x)$ has an attracting orbit of period 3; the points $x_0, x_1, \ldots, x_{49}$ have been suppressed in this graph.

◆ What is Chaos?

So, you're probably asking, what is chaos? The definition of chaos is a bit slippery. In fact, mathematicians are still arguing about a proper definition. But to get the idea across we'll use one due to Devaney.

Let S be a set such that if x lies in S, then $f^{\circ n}(x)$ belongs to S for all positive integers n. The set S is called *invariant*. If you start in an invariant set, you can't get out! Now let's define what we mean by chaos in an invariant set S.

A map $f : S \rightarrow S$ is *chaotic* if:

1. periodic points are dense in S;

2. f displays sensitive dependence on initial conditions in S; and

3. f is topologically transitive in S.

☞ A set U of real numbers is *open* if every point p of U has the property that all points in some interval $(p - a, p + a)$ are also in U.

The first condition of this definition is explained like this: A set A is *dense* in another set B if for every point x in B and every open set U containing x there exists points of A that are also in U. Therefore, condition 1 says that periodic points are almost everywhere in S. This means that S contains many periodic points; Theorem 13.1 gives a condition guaranteeing infinitely many of these points.

☞ Module 13 has examples.   In the second condition, *sensitive dependence on initial conditions* means that points that are initially close to one another eventually get moved far apart under iteration by f.

Finally, f is *topologically transitive* (or *mixing*) if given any pair of open sets U and V in S, some iterate of f takes one or more points of U into V. This means that points of open sets get spread throughout the set S.

The most significant item on this list for applied problems is sensitive dependence on initial conditions. Let's consider the logistic growth model at a parameter value where the dynamics are chaotic. Sensitive dependence implies that no matter how close two populations may be today, there will be a time in the future when the populations differ significantly. So environmental disturbances that cause small population changes will eventually lead to large changes, if chaotic dynamics exist.

Chaotic dynamics occur in a wide range of models. Although the definitions above are given in terms of a single scalar dynamical system, everything extends to higher dimensions, and many of the applications are two or three dimensional. In addition to models of population dynamics, chaos has been observed in models of the weather, electrical circuits, fluid dynamics, planetary motion, and many other phenomena. The relatively recent understanding of chaos has shed new light on the complexity and beauty of the world we inhabit.

◆ Complex Numbers and Functions

Probably the most popular type of discrete dynamical system is a *complex dynamical system* where the variables are complex numbers instead of real numbers. The intricate fractal structures common to images generated using complex dynamics have appeared everywhere from calendars to art shows and have inspired both artists and scientists alike. Many of the fundamental ideas of complex dynamics are identical to those of real dynamics and have been discussed in previous sections. In what follows, we will highlight both the similarities and differences between real and complex dynamics.

Recall that complex numbers arise when factoring quadratic polynomials with negative discriminant. Because the discriminant is negative we must take the square root of a negative real number, which we do by defining i to be $\sqrt{-1}$. We then write the *complex number* as $z = x + iy$. We say that x is the *real part of z* and y is the *imaginary part of z*. The complex number z is represented graphically on the complex plane by the point having coordinates (x, y). It is often useful to represent complex numbers in polar coordinates by letting $x = r\cos\theta$ and $y = r\sin\theta$ so that

$$z = r(\cos\theta + i\sin\theta) = re^{i\theta}$$

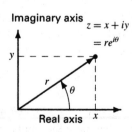

The remarkable relationship $\cos\theta + i\sin\theta = e^{i\theta}$ between polar coordinates and exponential functions is known as *Euler's Formula*. The number $r = \sqrt{x^2 + y^2}$ is the distance from the origin to the point z in the complex plane

☞ Euler's formula is also used in Chapter 4.

and is sometimes called the *modulus* of z, and it is denoted by $r = |z|$. The angle θ is called the *argument* of z. Note that the usual properties of exponential functions hold in the complex plane. Thus, given two complex numbers $z = re^{i\theta}$ and $w = se^{i\phi}$, their product is

$$zw = rse^{i(\theta+\phi)}$$

A *complex function* $f(z)$ takes a complex number z as its argument and returns a complex number $w = f(z)$. Differentiation proceeds as in the real case; for example, $(z^3)' = 3z^2$. Unlike functions of one real variable, we cannot graph a complex function since both the domain and range are two-dimensional.

◆ Iterating a Complex Function

Iteration of a complex function is identical to the iteration of a real function. Given an initial z-value z_0, iteration generates a sequence of complex numbers $z_1 = f(z_0)$, $z_2 = f(z_1)$, etc. Fixed and periodic points are defined in the same way as for real functions, as are stability and instability. Here are the previous criteria for stability, but now applied to complex functions.

- A fixed point z^* is stable if $|f'(z^*)| < 1$, and unstable if $|f'(z^*)| > 1$.
- A period-n point z^* (and its orbit) is stable if $|(f^{\circ n})'(z^*)| < 1$, and unstable if $|(f^{\circ n})'(z^*)| > 1$.

Let's consider a simple example to illustrate these ideas. Let $f(z) = z^2$. Then $z^* = 0$ is an attracting fixed point since $f(0) = 0$ and $|f'(0)| = 0$. If z is any point such that $|z| < 1$, then the sequence $\{f^{\circ n}(z)\}_{n=0}^{\infty}$ converges to 0 as $n \to \infty$. On the other hand, if $|z| > 1$, then the sequence $\{f^{\circ n}(z)\}_{n=0}^{\infty}$ goes to infinity as $n \to \infty$. To see what happens to values of z having modulus equal to 1, let's write $z = e^{i\theta}$. Then $f(z) = e^{2i\theta}$, which also has modulus 1. Thus all iterates of points on the unit circle $|z| = 1$ stay on the unit circle. The point $z^* = 1$ is a repelling fixed point since $f(1) = 1$ and $|f'(1)| = 2$. The period-2 points are found by solving $f(f(z)) = z^4 = z$. We can rewrite this equation as

$$z(z^3 - 1) = 0$$

One solution to this equation is $z^* = 0$, corresponding to the attracting fixed point, and another solution is $z^* = 1$, corresponding to the repelling fixed point. Notice that the fixed points of $f(z)$ remain fixed points of $f(f(z))$, or equivalently, are also period-2 points of $f(z)$. To find the other two solutions, we write $z = e^{i\theta}$ to get the equation

☞ Any period-n point is also a periodic point of all periods which are positive integer multiples of n.

$$e^{3i\theta} = 1$$

which we need to solve for θ. Since we are working in polar coordinates, we note that $1 = e^{i2n\pi}$ where n is an integer. This implies that $3\theta = 2n\pi$ and from this we find a second pair of period-2 points at $z = e^{2\pi i/3}$ and $z = e^{4\pi i/3}$. Both of these are repelling.

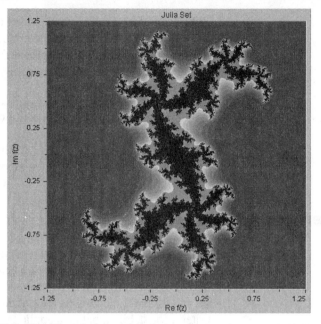

Figure 13.5: The filled Julia set for $f(z) = z^2 + c$, where $c = 0.4012 - 0.3245i$

✓   Show that $e^{2\pi i/3}$ and $e^{4\pi i/3}$ are repelling period-2 points of $f = z^2$. Show that $f^{\circ n}(z) \to 0$ as $n \to \infty$ if $|z| < 1$, and that $|f^{\circ n}(z)| \to \infty$ if $|z| > 1$. What is the "basin of attraction" of the fixed point $z = 0$?

◆ Julia Sets, the Mandelbrot Set, and Cantor Dust

The set of repelling periodic points of the function $f = z^2$ is dense on the unit circle, although we don't show that here. This leads us to the definition of the Julia set.

☞ The *closure* of a set A consists of the points of A together with all points that are limits of sequences of the points of A.

DEFINITION   The *filled Julia set K* of a complex-valued function f is the set of all points whose iterates remain bounded. The *Julia set J* of f is the closure of the set of repelling periodic points.

For $f = z^2$, the filled Julia set K of f is the set of all complex numbers z with $|z| \leq 1$, while the Julia set J of f is the unit circle $|z| = 1$. This is a very simple example of a Julia set. In general, Julia sets are highly complicated objects having a very intricate fractal structure. For example, see Figure 13.5 and Screens 3.3 and 3.4 of Module 13.

In the above example, the Julia set J divides those points that iterate to infinity (points outside the unit circle) and those that converge to the attracting fixed point (points inside the unit circle). This division of the domain by the Julia set is often the case in complex dynamics and provides a way of

☞ In the Discrete Tool of ODE Architect, the coloring is reversed. Points in the Julia set are colored black and points whose orbits diverge past the predetermined bound are colored with various colors according to their divergence rates (e.g., red is the fastest, dark blue the slowest).

☞ A complex function f is *analytic* if its derivatives of every order exist. A point $\tilde{z}$ is a *critical point* of f if $f'(\tilde{z}) = 0$.

numerically computing the filled Julia set of a given function f. Assign a complex number to each screen pixel. Then use each pixel (i.e., complex number) as an initial condition and iterate to determine whether the orbit of that point exceeds some predetermined bound (for example $|z| = 50$). If it does, we say the orbit diverges and we color the point black. If not, we color the point red to indicate it is in the filled Julia set.

Earlier in this chapter we saw the importance of attracting periodic orbits in building a bifurcation diagram for a real map f. Although we didn't mention it then, we can home in on an attracting periodic orbit of f (if there is one) by starting at $x_0 = \tilde{x}$ if $f'(x)$ is zero at $\tilde{x}$ and nowhere else. Complex functions $f(z)$ for which $f'(\tilde{z}) = 0$ at exactly one point $\tilde{z}$ have the same property as the following theorem shows.

THEOREM 13.3 Let f be an analytic complex-valued function with a unique critical point $\tilde{z}$. If f has an attracting periodic orbit, then the forward orbit of $\tilde{z}$ converges to this orbit.

Let's look at some of the implications of this theorem with the family of functions $f_c(z) = z^2 + c$ where $c = a + ib$ is a complex parameter. For each value of c the only critical point is $\tilde{z} = 0$. To find an attracting periodic orbit for a given value of c we need to compute the orbit

$$\{0, c, c^2 + c, \dots\}$$

and see if the orbit converges or not. If it does, we found the attracting periodic orbit; if not, there doesn't exist one. Let's see what happens when we set $c = 1$ to give the function $f_1(z) = z^2 + 1$. The orbit of the critical point is $\{0, 1, 2, 5, 26, \dots\}$, which goes to infinity. Thus, f_1 has *no* attracting periodic orbit and the Julia set does not divide points that converge to a periodic orbit from points that iterate to infinity. In fact, it can be shown that this Julia set is totally disconnected; it is sometimes referred to as *Cantor dust*. Click on outlying points on the edge of the Mandelbrot (defined below) set in Screen 3.5 of Module 13 and you will generate Cantor dust in the upper graphics screen.

This leads to another question. If some functions in the family f_c have connected Julia sets (such as $f_0 = z^2$) and other functions in the family have totally disconnected Julia sets (such as f_1), what set of points in the c plane separates these distinctive features? This set is the *boundary* of the Mandelbrot set. The *Mandelbrot set M* of the function $f_c(z) = z^2 + c$ is defined as the set of all complex numbers c such that the orbit $\{f_c^{\circ n}(0)_{n=1}^{\infty}\}$ remains bounded, that is, $|f_c^{\circ n}(0)| \le K$ for some positive number K and all integers $n \ge 0$.

This definition leads us to an algorithm for computing the Mandelbrot set M. Assign to each pixel a complex number c. Choose a maximum number of iterations N and determine whether $|f_c^{\circ n}(0)| < 2$ for all $n \le N$ (it can be proven that if $|f_c^{\circ n}(0)| > 2$ for some n, then the orbit goes to infinity). If so, then color this point green to indicate that it is in the Mandelbrot set. Otherwise, color this point black. It is this computation that gives the wonderfully intricate Mandelbrot set; see Figure 13.6 and Screens 3.4 and 3.5 of Module 13.

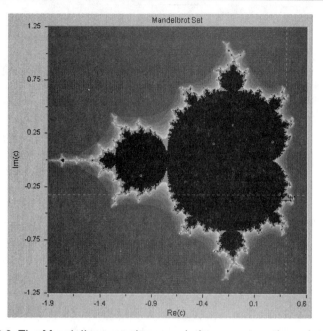

Figure 13.6: The Mandelbrot set; the cross-hairs are set on the point $0.4012 - 0.3245i$, which gives the Julia set shown in Figure 13.5.

The Mandelbrot set actually contains much more information than is described here. It is, in fact, the bifurcation diagram for the family of functions $f_c(z) = z^2 + c$. Each "blob" of the set corresponds to an attracting periodic orbit of a particular period. Values of c in the big cardioid shown on Screen 3.4 of Module 13 give attracting period-1 orbits for f_c. Values of c in the circle immediately to the left of this cardioid give attracting period-2 orbits for f_c. Other "blobs" give other attracting periodic behaviors.

Although we have only defined the Mandelbrot set for the specific family $f_c = z^2 + c$, it can be defined in an anlagous way for other families of complex functions (see the Discrete Tool). One final note on Julia sets and the Mandelbrot set. You've probably seen intricately colored versions of these objects on posters or elsewhere. The coloring is usually determined by how "fast" orbits tend to infinity. The color scheme is, of course, up to the programmer.

Module 13 introduces and lets you play with three important discrete dynamical systems—linear, logistic, and a third that uses complex numbers. Explorations 13.1–13.4 extend these ideas and introduce other maps with curious behavior under iteration.

References Alligood, K.T., Sauer, T.D. and Yorke, J.A., *Chaos: An Introduction To Dynamical Systems*, (1997: Springer-Verlag). "Period Three Implies Chaos" and Šarkovskii's theorem are described in the third of thirteen chapters in the marvelous book.

Bollt, E.M. "Controlling the Chaotic Logistic Map" in *Primus* (March, 1997) pp. 1–18. A lovely undergraduate project, first written as Project 3 of MA385 at the U.S. Military Academy, Dept. of Mathematical Sciences, West Point, NY 10928.

Devaney, R.L., *A First Course in Chaotic Dynamical Systems, Theory and Experiment*, (1992: Addison Wesley). This is the "intermediate" book by Devaney on chaotic dynamical systems. Chapter 11 concerns Šarkovskii's Theorem, and Yorke and Li's theorem that "Period 3 Implies Chaos".

Doebeli, M. and Ruxton, G. "Controlling Spatial Chaos in Metapopulations with Long-Range Dispersal" in *Bulletin of Mathematical Biology*, Vol. 59 (1997) pp. 497–515. Emphasizes the frequent use of the linear map and the logistic map in population dynamics.

Guimez, J. and Matias, M.A. "Control of Chaos in Unidimensional Maps" in *Physics Letters A* Vol. 181 (September 1993) pp. 29–32

Peitgen, H.-O. and Richter, P.H., *The Beauty of Fractals*, (1986: Springer-Verlag)

Saha, P. and Strogatz, S.H. "The Birth of Period 3" in *Mathematics Magazine* Vol. 68 (1995) pp. 42–47

Strogatz, S., *Nonlinear Dynamics and Chaos*, (1994: Addison-Wesley). A very readable text, mostly on ODEs, but Chapter 10 on One-Dimensional Maps is particularly relevant to this chapter/module of ODE Architect.

Yorke, J. and Li, T.Y., "Period 3 Implies Chaos" in the *American Mathematical Monthly* 82 (1975) pp. 985–992

Here are three sources on chaos that are less technical and more accessible, but still give an accurate description of what chaos is all about.

Peak, D. and Frame, M., *Chaos under Control*, (1994: W.H. Freeman)

Stewart, I., *Does God Play Dice?*, (1990, Blackwell)

Stoppard, T., the play "Arcadia", (1993: Faber and Faber). Yes! A play about chaos.

Answer questions in the space provided, or on
attached sheets with carefully labeled graphs. A
notepad report using the Architect is OK, too.

Name/Date _____

Course/Section _____

Exploration 13.1. One-Dimensional Maps and the Discrete Tool

1. Go to the Discrete Tool and enter the proportional growth model $x_{n+1} = cx_n$, where c is the parameter. For the range $0 \leq n \leq 30$ and the initial condition $x_0 = 0.5$, explore and describe what happens to the iteration map, time series, and bifurcation diagram as the parameter is increased from -2 to 2. For what values of c is there a sudden change in the behavior of the iterates (the bifurcation values of c)? For what values of c are there 1, 2, or infinitely many fixed or periodic points? Which of these points are attractors? Repellers?

2. Go to the Discrete Tool and explore and describe what happens to the iteration map, the time series, and the bifurcation diagram as the parameter c for the logistic map $g_c(x) = cx(1 - x)$ is incremented from 1 to 4. Use the range $50 \leq n \leq 150$ to avoid an initial wandering, and the initial condition $x_0 = 0.5$. Describe what all three graphs on the tool screen look like at values of c where there is a periodic orbit. What is the period? Go as far forward as you can with the period-doubling sequence of values of c: 3, 3.434, ... What are the corresponding periods? [*Suggestion:* Zoom on the bifurcation diagram.] Repeat with the sequence 3.83, 3.842, ...

3. In the Discrete Tool enter the tent map T_c on the interval $0 \leq x \leq 1$:

$$T_c(x) = c(1 - 2\,\mathrm{abs}(x - 0.5)) = \begin{cases} 2cx, & 0 \leq x \leq 0.5 \\ 2c(1 - x), & 0.5 \leq x \leq 1 \end{cases}$$

where the parameter c is allowed to range from 0 to 1. Describe and explain what you see as c is incremented from 0 to 1. [*Suggestion:* use the Edit option in the Menu box for the Bifurcation Diagram and set $200 \leq n \leq 300$ in order to suppress the initial transients.] Any orbits of period 2? Period 3?

Answer questions in the space provided, or on
attached sheets with carefully labeled graphs. A
notepad report using the Architect is OK, too.

Name/Date _____

Course/Section _____

Exploration 13.2. Circle Maps

Another common type of discrete dynamical system is a *circle map*, which maps the perimeter of the unit circle onto itself. These functions arise when modeling coupled oscillators, such as pendulums or neurons. The simplest types of circle maps are rotations that take the form

$$R_\omega(\theta) = (\theta + \omega) \bmod 2\pi$$

where $0 \le \theta \le 2\pi$ and ω is a constant.

1. Show that if $\omega = (p/q)\pi$ with p and q positive integers and p/q in lowest terms, then every point has period q.

2. Show that if $\omega = a\pi$ with a an irrational number, then no point on the circle is periodic.

3. What is the long-term behavior of the orbit of a point on the circle if $\omega = a\pi$, where a is an irrational number?

Answer questions in the space provided, or on attached sheets with carefully labeled graphs. A notepad report using the Architect is OK, too.

Name/Date _____

Course/Section _____

Exploration 13.3. Two-Dimensional Maps and the Discrete Tool

A two-dimensional discrete dynamical system looks like this:

$$x_{n+1} = f(x_n, y_n, c)$$
$$y_{n+1} = g(x_n, y_n, c) \tag{3}$$

where f and g are given functions and c is a "place holder" for parameters. For given values of c, x_0, and y_0, system (3) defines an orbit of points

$$(x_0, y_0), \quad (x_1, y_1), \quad (x_2, y_2), \ldots$$

in the xy-plane. The-two dimensional tab in the Discrete Tool allows you to explore discrete systems of the form of system (3).

1. Open the Discrete Tool and explore the default system (a version of what is known as the *Hènon Map*):

$$x_{n+1} = 1 + y_n - ax_n^2$$
$$y_{n+1} = bx_n \tag{4}$$

where a and b are parameters. For fixed values of the parameters a and b find the fixed points. Are they sinks, sources, or neither? How sensitive is the long-term behavior of an orbit to small changes in the initial point (x_0, y_0)? What happens if you increment a through a range of values? If you increment b? Any period-doubling sequences? In your judgment, is there any long-term chaotic wandering? [*Suggestion:* Keep the values of a and b within small ranges of their default values to avoid instabilities.]

2. Repeat Problem 1 with the following version of the Hènon map:

$$x_{n+1} = a - x_n^2 + by_n$$
$$y_{n+1} = x_n$$

Start with $a = 1.28$, $b = -0.3$, $x_0 = 0$, $y_0 = 0$.

Answer questions in the space provided, or on
attached sheets with carefully labeled graphs. A
notepad report using the Architect is OK, too.

Name/Date _____

Course/Section _____

Exploration 13.4. Julia and Mandelbrot Sets and the Discrete Tool

Note that the color schemes for the Julia and Mandelbrot sets in Module 13 differ from those in the discrete tool.

1. Use the Discrete Tool to explore the Mandelbrot set and Julia sets for the complex family $f_c = z^2 + c$. What happens to the filled Julia sets as you move c from inside the Mandelbrot set up toward the boundary, then across the boundary and out beyond the Mandelbrot set? Describe how the Julia sets change as you "walk" along the edge of the Mandelbrot set.

2. Repeat Problem 1 for the complex family $g_c = c \sin z$.

3. Repeat Problem 1 for the family $h_c = ce^z$.

GLOSSARY

Acceleration The acceleration of a moving body whose position at time t is $u(t)$ is given by

$$\frac{d^2 u}{dt^2}$$

Air resistance A body moving through air (or some other medium) is slowed down by a resistive force (also called a drag or damping force) that acts opposite to the body's velocity. See also "Viscous damping" and "Newtonian damping."

Amplitude The amplitude of a periodic oscillating function $u(t)$ is half the difference between its maximum and minimum values.

Angular momentum The angular momentum vector of a body rotating about an axis is its moment of inertia about the axis times its angular velocity vector.

This is the analog in rotational mechanics of momentum (mass times velocity) in linear mechanics.

Angular velocity An angular velocity vector, $\omega(t)$, is the key to the relation between rotating body axes and a fixed coordinate system of the observer. The component ω_j of the vector $\omega(t)$ along the jth body axis describes the spin rate of the body about that axis.

Autocatalator This is a chemical reaction of several steps, at least one of which is autocatalytic.

Autocatalytic reaction In an autocatalytic reaction, a chemical species stimulates more of its own production than is destroyed in the process.

Autonomous ODE An autonomous ODE has no explicit mention of the independent variable (usually t) in the rate equations. For example, $x' = x^2$ is autonomous, but $x' = x^2 + t$ is not.

Balance law The balance law states that the net rate of change of the amount of a substance in a compartment equals the net rate of flow in minus the net rate of flow out.

Beats When two sinusoids of nearly equal frequencies are added the result appears to be a high frequency sinusoid modulated by a low frequency sinusoid called a beat. A simple example is given by the function $(\sin t)(\sin 10t)$, where the first sine produces an "amplitude modulation" of the second.

Bessel functions of the 1st kind The Bessel function of the first kind of order zero,

$$J_0(s) = 1 - \frac{1}{4}s^2 + \cdots + (-1)^n \frac{s^{2n}}{n!^2 2^{2n}} + \cdots$$

is a solution of Bessel's equation of order zero, and is bounded and convergent for all s.

Bessel functions of the 2nd kind The Bessel function of the second kind of order zero, $Y_0(s)$, is another solution of Bessel's equation of order zero. It is much more complicated than $J_0(s)$, and

$$Y_0(s) \to \infty \quad \text{as} \quad s \to 0+$$

See Chapter 11 for a complete formula for $Y_0(s)$ that involves a logarithmic term, $J_0(s)$, and a complicated (but convergent) infinite series.

Bessel's equation Bessel's equation of order $p \geq 0$ is

$$s^2 w''(s) + s w'(s) + (s^2 - p^2)w = 0$$

where p is a nonnegative constant. Module 11 considers only $p = 0$. See Chapter 11 for $p > 0$.

Bessel's equation, general solution of Bessel's equation of order zero is second-order and linear. The general solution is the set of all linear combinations of $J_0(s)$ and $Y_0(s)$.

Bifurcation diagram A bifurcation diagram describes how the behavior of a dynamical system changes as a parameter varies. It can appear in studies of iteration or of differential equations.

In the case of a single real parameter, a bifurcation diagram plots a parameter versus something indicative of the behavior, such as the variable being iterated (as in Module 13, Nonlinear Behavior) or a single variable marking location and stability of equilibrium points for a differential equation.

In iteration of a function of a complex variable, two dimensions are needed just to show the parameter, but different colors can be used to show different behaviors (as in Module 13, Complex Dynamics).

Cantor Set, Cantor Dust A Cantor set was first detailed by Henry Smith in 1875, but was named in honor of Georg Cantor, the founder of set theory, after he used this bizarre construction in 1883. Now Cantor sets are found in many guises in discrete dynamical systems.

A Cantor set is a totally disconnected set, in a finite space, with uncountably many points. A typical construction is to delete a band across the middle of a set, then to delete the middle of both pieces that are left, and then to repeat this process indefinitely.

Julia sets (see glossary) for parameter values outside the Mandelbrot set (see glossary) are Cantor dusts, constructed by a similar algorithm. See Companion Book for References.

Carrying capacity The carrying capacity K of an environment is the maximum number of individuals that the environment can support at steady state. If there are fewer individuals than the carrying capacity in the environment, the population will grow; if there are more individuals, the population will decline.

A widely used model for population dynamics involving a carrying capacity is the logisitc ODE

$$\frac{dN}{dt} = rN(1 - N/K)$$

where r is the intrinsic growth rate constant.

Cascade A cascade is a compartment model where the "flow" through the compartments is all one direction.

Center A center is an equilibrium point of an autonomous planar linear system for which the eigenvalues are conjugate imaginaries $\pm i\beta$, $\beta \neq 0$. All nonconstant orbits of an autonomous planar linear system with a center are simple closed curves enclosing the equilibrium.

Centering an equilibrium If p^* is an equilibrium point of the system $x' = f(x)$ (so $f(p^*) = 0$), then the change of coordinates $x = y + p^*$ moves p^* to the origin in the y-coordinate system.

Chain rule The chain rule for differentiating a function $L(\theta(t), y(t))$ with respect to t is

$$\begin{aligned}
\frac{dL}{dt} &= \frac{\partial L}{\partial \theta}\frac{d\theta}{dt} + \frac{\partial L}{\partial y}\frac{dy}{dt} \\
&= L_\theta \theta' + L_y y' \\
&= L_\theta y + L_y y'
\end{aligned}$$

Chaos Mathematical chaos is a technical term that describes certain nonperiodic behavior of a discrete dynamical system (Module 13) or solutions to a differential equation (Module 12). A system is said to be chaotic in a region if all of the following are true.

- It exhibits sensitive dependence on initial conditions.
- Periodic unstable orbits occur almost everywhere.
- Iterates of intervals get "mixed up."

Chaotic behavior never repeats, revisits every neighborhood infinitely often, but is not random. Each step is completely determined by the previous step.

An equivalent list of requirements appears in Module 12, Screen 1.4. Further discussion appears in Chapter 13.

Characteristic equation The characteristic equation of a square matrix A is $\det(\lambda I - A) = |\lambda I - A| = 0$. For a 2×2 matrix, this reduces to $\lambda^2 - \operatorname{tr} A\lambda + \det A = 0$ whose solutions, called eigenvalues of A are

$$\lambda = \frac{\operatorname{tr} A \pm \sqrt{\operatorname{tr}^2 A - 4\det A}}{2}$$

Chemical law of mass action The rate of a reaction step is proportional to the product of the concentrations of the reactants.

Example: If one unit of species X produces one unit of product Y in a reaction step, the rate of the step is kx, where k is a positive constant. Thus, we have

$$x' = -kx, \qquad y' = kx$$

Example (Autocatalysis): If one unit of species X reacts with two units of Y and produces three units of Y in an autocatalytic step, the reaction rate is

$$axyy = axy^2$$

where a is a positive constant. Thus, we have

$$x' = -axy^2, \qquad y' = 3axy^2 2 - 2axy^2 = axy^2$$

because one unit of X is destroyed, while three units of Y are created, and two are consumed.

Combustion model The changing concentration $y(t)$ of a reactant in a combustion process is modeled by the IVP

$$y' = y^2(1 - y), \qquad y(0) = a, \quad 0 \leq t \leq 2/a$$

where a is a small positive number that represents a disturbance from the pre-ignition state $y = 0$. R. E. O'Malley studied the problem in his book, *Singular Perturbation Methods for Ordinary Differential Equations*, (1991: Springer).

Compartment model A compartment model is a set of boxes (the compartments) and arrows that shows the flow of a substance into and out of the different boxes.

Component graphs A component graph of a solution of a differential system is a graph of one of the dependent variables as a function of t.

Example: For the ODE system

$$x' = F(x, y)$$
$$y' = G(x, y)$$

the component graphs are the plots of a solution $x = x(t)$ and $y = y(t)$ in the respective tx- and ty-planes.

Concentration The concentration of a substance is the amount of the substance dissolved per unit volume of solution.

Connected set A connected set is a set with no islands. In the early 1980's Adrien Douady (Université Paris XI, Orsay and Ecole Normale Supérieure) and John Hubbard (Cornell University) proved that the Mandelbrot set (see Glossary) was connected. They did this by showing that its exterior could be put in a one-to-one correspondence with the exterior of a disk. They found in the process that all the angles one might note while walking around the boundary of the disk have special analogs on the Mandelbrot set. Halfway around the disk from the rightmost point corresponds to being at the tip of the Mandelbrot set, while one third or two thirds the way around the disk corresponds to the "neck" where the biggest ball attaches to the cardioid.

Conserved quantity A function $E(q, y)$ is conserved along a trajectory $q = q(t)$, $y = y(t)$, of a system $q' = f(q, y)$, $y' = g(q, y)$, if $dE(q(t), y(t))/dt = 0$.

As time changes, the value of E stays constant on each trajectory, although the value will vary from one trajectory to another. The graph of each trajectory in the qy-phase plane lies on one of the level sets $E =$ constant.

This idea of a conserved quantity can be extended to any autonomous system of ODEs. An autonomous system is conservative if there is a function E that stays constant along each trajectory, but is nonconstant on every region (i.e., varies from trajectory to trajectory).

Cycle In a discrete dynamical system, including a Poincaré section, a cycle is a sequence of iterates that repeats. The number of iterates in a cycle is its period.

For an autonomous differential system, a cycle is a nonconstant solution $x(t)$ such that $x(t + T) = x(t)$, for all t, where T is a positive constant. The smallest value of T for a cycle is its period.

For a cycle in a system of 2 ODEs, see Limit Cycle.

Damped pendulum A real pendulum of length L is affected by friction or air resistance that is a function of L, θ, and θ', and acts opposite to the direction of motion.

Throughout the Linear and Nonlinear Pendulums submodule of Module 10, we assume that, if there is any damping, it is viscous (see Viscous damping); i.e., the damping force is given by $-bL\theta'$. The minus sign tells us that damping acts opposite to the velocity.

Module 4 makes a more detailed study of the effects of damping on a linear oscillator, as does Module 11 for the spring in the Robot and Egg.

Damping Damping can arise from several sources, including air resistance and friction. The most common model of damping is viscous damping—the damping force is assumed to be proportional to the velocity and acts opposite to the direction of motion. See also Newtonian damping.

Dense orbit An orbit $x(t)$ of a system of ODEs $x' = f(t, x)$ is dense in a region R of x-space if the orbit gets arbitrarily close to every point of R as time goes on.

That is, if x_1 is any point in R, and ε is any positive number, then, at some time t_1, the distance between $x(t_1)$ and x_1 is less than ε.

Determinant The determinant of the 2×2 matrix

$$A = \begin{bmatrix} a & b \\ c & d \end{bmatrix}$$

is det $A = ad - bc$.

Deterministic A system of ODEs is said to be deterministic if the state of the system at time t is uniquely determined by the state of the system at the initial time.

For example, the single first-order ODE $x' = f(t, x)$ is deterministic if f and $\partial f/\partial x$ are continuous functions of t and x, and for each set of initial data (t_0, x_0) there is exactly one solution $x(t)$.

Thus, if you were to choose the same initial data a second time and watch the solution curve trace out in time again, you would see exactly the same curve.

Dimensionless variables Suppose that a variable x is measured in units of kilograms and that x varies from 10 to 500 kilograms. If we set $y = (x \text{ kilograms})/(100 \text{ kilograms})$, y is dimensionless, and $0.1 \le y \le 5$. The smaller range of values is useful for computing. The fact that y has no units is useful because it no longer matters if the units are kilograms, grams, or some other units.

When variables are scaled to dimensionless quantities, they are typically divided by a constant somewhere around the middle of the expected range of values. For example, by dividing a chemical concentration by a "typical" concentration, we obtain a dimensionless concentration variable. Similarly, dimensionless time is obtained by dividing ordinary time by a "standard" time.

Direction field A direction field is a collection of line segments which shows the slope of the trajectories for an autonomous ODE system

$$\frac{dx}{dt} = F(x, y)$$

$$\frac{dy}{dt} = G(x, y)$$

at a representative grid of points. An arrowhead on a segment shows the direction of motion.

Disconnected Julia set A disconnected Julia set is actually a Cantor dust. It is composed entirely of totally disconnected points, which means that it is almost never possible to land on a point in the Julia set by clicking on a pixel. You will probably find that every click you can make starts an iteration that goes to infinity, only because you cannot actually land on an exact enough value to show a stable iteration.

Discrete dynamical system A discrete dynamical system takes the form $u_{n+1} = f(u_n)$, where the variable u_n gives the state of the system at "time" n, and u_{n+1} is the state of the system at time $n + 1$. See Module 13.

Eigenvalues The eigenvalues of a matrix A are the numbers λ for which

$$Av = \lambda v$$

for some nonzero vector v (the vector v is called an eigenvector). The eigenvalues λ of a 2×2 matrix A are the solutions to the characteristic equation of A:

$$\lambda^2 - \operatorname{tr} A\lambda + \det A = 0$$

$$\lambda = \frac{\operatorname{tr} A \pm \sqrt{(\operatorname{tr} A)^2 - 4\det A}}{2}$$

where $\operatorname{tr} A$ is the trace of A, and $\det A$ is the determinant of A. If a linear or linearized system of ODEs is $z' = A(z - p^*)$, and if the real parts of the eigenvalues of A are positive, then trajectories flow away from the equilibrium point, p^*. If the real parts are negative, then trajectories flow toward p^*.

Eigenvector An eigenvector of a matrix A is a nonzero vector, v, that satisfies $Av = \lambda v$ for some eigenvalue λ. The ODE Architect Tool calculates eigenvalues and eigenvectors of Jacobian matrices at any equilibrium point of an autonomous system (linear or nonlinear).

Eigenvectors play a strong role in the local geometry of phase portraits at an equilibrium point.

Energy In physics and engineering, energy is defined by

$$E = \text{kinetic energy} + \text{potential energy}$$

where kinetic energy is interpreted to be the energy of motion, and the potential energy is the energy due to some external force, such as gravity, or (in electricity) a battery, or a magnet. If energy is conserved, i.e., stays at a constant level, then the system is said to be conservative.

If we are dealing with the autonomous differential system

$$x' = y, \qquad y' = -v(x) \tag{5}$$

we can define an "energy function" by

$$E = \frac{1}{2}y^2 + V(x)$$

where $dV/dx = v(x)$. Note that E is constant along each trajectory, because $dE/dt = y\,dy/dt + (dV/dx)(dx/dt) = y(-v(x)) + v(x)(y) = 0$, where the ODEs in system (5) have been used. The term $(1/2)y^2$ is the "kinetic energy". $V(x)$ is the "potential energy" in this context. See Chapter 10 for more on these ideas.

Epidemic An epidemic occurs in an epidemiological model if the number of infectives, $I(t)$, increases above its initial value, I_0. Thus, an epidemic occurs if $I'(0) > 0$.

Equilibrium point An equilibrium point p^* in phase (or state) space of an autonomous ODE, is a point at which all derivatives of the state variables are zero—a stationary point—a steady-state value of the state variables. For example, for the autonomous system,

$$x' = F(x, y), \qquad y' = G(x, y)$$

if $F(x^*, y^*) = 0$, $G(x^*, y^*) = 0$, then $p^* = (x^*, y^*)$ is an equilibrium point, and $x = x^*$, $y = y^*$ (for all t) is a constant solution.

For a discrete dynamical system, an equilibrium point p^* is one for which $f(p^*) = p^*$, so that $p_{n+1}^* = p_n^*$, for all n; p^* is also called a fixed point of the system.

Estimated error For the solution $u(t)$ of the IVP $y' = f(t, y)$, $y(t_0) = y_0$, the local error at the nth step of the Euler approximation is given by

$$e_n = \text{Taylor series of } u(t) - \text{Euler approximation}$$

$$= \frac{1}{2}h^2 u''(t_n) + h^3 u'''(t_n) + \cdots$$

If the true solution, $u(t)$, is not known, we can approximate e_n for small h by

$$e_n \approx \text{Taylor approx.} - \text{Euler approx.} \approx \frac{1}{2}h^2 u''(t_n)$$

Euler's method Look at the IVP $y' = f(t, y)$, $y(t_0) = y_0$. Euler's method approximates the solution $y(t)$ at discrete t values. For step size h, put $t_{n+1} = t_n + h$ for $n = 0, 1, 2, \ldots$. Euler's method approximates

$$y(t_1), y(t_2), \ldots$$

by the values

$$y_1, y_2, \ldots$$

where

$$y_{n+1} = y_n + hf(t_n, y_n), \quad \text{for } n = 0, 1, 2, \ldots$$

Existence and uniqueness A basic uniqueness and existence theorem says that, for the IVP,

$$x' = F(x, y, t), \qquad y' = G(x, y, t),$$
$$x(t_0) = x_0, \qquad y(t_0) = y_0$$

a unique solution $x(t)$, $y(t)$ exists if F, G, $\partial F/\partial x$, $\partial F/\partial y$, $\partial G/\partial x$, and $\partial G/\partial y$ are all continuous in some region containing (x_0, y_0).

Fixed point A fixed point, p^*, of a discrete dynamical system is a point for which $x_{n+1} = f(x_n) = x_n$. That is, iteration of such a point simply gives the same point.

A fixed point can also be called an equilibrium or a steady state. A fixed point may be a sink, a source, or a saddle, depending on the character of the eigenvalues of the associated linearization matrix of the iterating function.

Forced damped pendulum A forced, viscously damped pendulum has the modeling equation

$$mx'' + bx' + k\sin x = F(t)$$

The beginning of Module 10 explains the terms and parameters of this equation using θ instead of x. Module 12 examines a case where chaos can result, with $b = 0.1$, $m = 1$, $k = 1$, $A = 1$, and $F(t) = \cos t$. All three submodules of Module 12 are involved in explaining the behaviors, and the introduction to the Tangled Basins submodule shows a movie of what happens when b is varied from 0 to 0.5.

Forced pendulum Some of the most complex and curious behavior occurs when the pendulum is driven by an external force. In Module 10, The Pendulum and Its Friends, you can experiment with three kinds of forces in the Linear and Nonlinear Pendulums submodule, and an internal pumping force in the Child on a Swing submodule. But, for truly strange behavior, take a look at Module 12, Chaos and Control.

Fractal dimension Benoit Mandelbrot in the early 1980's coined the word "fractal" to apply to objects with dimensions between integers. The boundary of the Mandelbrot set (see glossary) is so complicated that its dimension is surely greater than one (the dimension of any "ordinary" curve). Just how much greater remained an open question until 1992 when the Japanese mathematician Mitsuhiro Shishikura proved it is actually dimension two!

Frequency The frequency of a function of period T is $1/T$. Another widely used term is "circular frequency", which is defined to be $2\pi/T$. For example, the periodic function $\sin(3t)$ has period $T = 2\pi/3$, frequency $3/(2\pi)$, and circular frequency 3.

General solution Consider the linear system $x' = Ax + u$ [where x has 2 components, A is a 2×2 matrix of constants, and u is a constant vector or a function only of t]. Let A have distinct eigenvalues λ_1, λ_2 with corresponding eigenvectors v_1, v_2. All solutions of the system are given by the so-called general solution:

$$x(t) = C_1 e^{\lambda_1 t} v_1 + C_2 e^{\lambda_2 t} v_2 + \tilde{x}$$

where $\tilde{x}$ is any one particular solution of the system and C_1 and C_2 are arbitrary constants.. If u is a constant vector, then $\tilde{x} = p^*$, the equilibrium of the system. If x has more than two dimensions, terms of the same form are added until all dimensions are covered. Note that, if $u = 0$, $p^* = 0$ is an equilibrium.

Geodesic Any smooth curve can be reparametrized to a unit speed curve $x(t)$, where $|x'(t)| = 1$. Unit-speed curves $x(t)$ on a surface are geodesics if the acceleration vector $x''(t)$ is perpendicular to the surface at each point $x(t)$.

It can be shown that a geodesic is locally length-minimizing, so, between any two points sufficiently close, the geodesic curve is the shortest path.

GI tract The gastro-intestinal (GI) tract consists of the stomach and the intestines.

Gravitational force The gravitational force is the force on a body due to gravity. If the body is near the earth's surface, the force has magnitude mg, where m is the body's mass, and the force acts downward. The value of acceleration due to gravity, g, is 32 ft/sec^2 (English units), 9.8 meters/sec^2 (metric units).

Great circle A great circle on a sphere is an example of a geodesic. You can test this with a ball and string. Hold one end of the string fixed on a ball. Choose another point some distance away, and find the geodesic or shortest path by pulling the string tight between the two points. You will find that it always is along a circle centered at the center of the ball, which is the definition of a great circle.

Hooke's law Robert Hooke, an English physicist in the seventeenth century, stated the law that a spring exerts a force, on an attached mass, which is proportional to the displacement of the mass from the equilibrium position and points back toward that position.

Initial condition An initial condition specifies the value of a state variable at some particular time, usually at $t = 0$.

Initial value problem An initial value problem (IVP) consists of a differential equation or a system of ODEs and an initial condition specifying the value of the state variables at some particular time, usually at $t = 0$.

Integral surfaces The surface S defined by $F(x, y, z) = C$, where C is a constant, is an integral surface of the autonomous system

$$x' = f(x, y, z), \qquad y' = g(x, y, z), \qquad z' = h(x, y, z)$$

if

$$\frac{d}{dt} F(x, y, z) = \frac{\partial F}{\partial x}\frac{dx}{dt} + \frac{\partial F}{\partial y}\frac{dy}{dt} + \frac{\partial F}{\partial z}\frac{dz}{dt}$$
$$= \frac{\partial F}{\partial x} f + \frac{\partial F}{\partial y} g + \frac{\partial F}{\partial z} h = 0$$

for all x, y, z. We get a family of integral surfaces by varying the constant C. An orbit of the system that touches an integral surface stays on it. The function F is called an integral of the system.

For example, the family of spheres

$$F = x^2 + y^2 + z^2 = \text{constant}$$

is a family of integral surfaces for the system

$$x' = y, \qquad y' = z - x, \qquad z' = -y$$

because

$$2xx' + 2yy' + 2zz' = 2xy + 2y(z - x) + 2z(-y) = 0$$

Each orbit lies on a sphere, and each sphere is covered with orbits.

Intermediate An intermediate is a chemical produced in the course of a reaction which then disappears as the reaction comes to an end.

Intrinsic growth rate At low population sizes, the net rate of growth is essentially proportional to population size, so that $N' = rN$. The constant r is called the intrinsic growth rate constant. It gives information about how fast the population is changing before resources become limited and reduce the growth rate.

Iteration Iteration generates a sequence of numbers by using a given number x_0 and the rule $x_{n+1} = f(x_n)$, where $f(x)$ is a given function. Sometimes, x_n is written as $x(n)$.

IVP See initial value problem.

Jacobian matrix The system $x' = F(x, y)$, $y' = G(x, y)$, has the Jacobian matrix

$$J = \begin{bmatrix} \frac{\partial F}{\partial x} & \frac{\partial F}{\partial y} \\ \frac{\partial G}{\partial x} & \frac{\partial G}{\partial y} \end{bmatrix}$$

The eigenvalues and eigenvectors of this matrix at an equilibrium point p^* help determine the local geometry of the phase portrait.

Jacobian matrices J can be defined for autonomous systems of ODEs with any number of state variables.

The ODE Architect Tool will find eigenvalues and eigenvectors of J at any equilibrium point.

Julia Set In complex dynamics, a Julia set for a given function $f(z)$ separates those points that iterate to infinity from those that do not. See the third submodule of Module 13 Dynamical Systems.

Julia sets were discovered about 1910 by two French mathematicians, Pierre Fatou and Gaston Julia. But, without computer graphics, they were unable to see the details of ragged structure that today display Cantor sets, self-similarity and fractal properties.

Kinetic energy of rotation The kinetic energy of rotation of a gyrating body is

$$E = \frac{1}{2}(I_1\omega_1^2 + I_2\omega_2^2 + I_3\omega_3^2)$$

where I_j and ω_j are, respectively, the moment of inertia and the angular velocity about the body axis, j, for $j = 1, 2, 3$.

Lift The lift force on a body moving through air is a force that acts in a direction orthogonal to the motion. Its magnitude may be modeled by a term which is proportional to the speed or to the square of the speed.

Limit cycle A cycle is a closed curve orbit of the system

$$x' = F(x, y)$$
$$y' = G(x, y)$$

A cycle is the orbit of a periodic solution.

An attracting limit cycle is a cycle that attracts all nearby orbits as time increases; a repelling limit cycle repels all nearby orbits as time increases.

Linearization For a nonlinear ODE, a linearization (or linear approximation) can be made about an equilibrium, $p^* = (x^*, y^*)$, as follows:

For $x' = F(x, y)$, $y' = G(x, y)$, the linearized system is $z' = J(z - p^*)$, where J is the Jacobian matrix evaluated at p^*, i.e.,

$$\begin{bmatrix} x \\ y \end{bmatrix}' = \begin{bmatrix} \frac{\partial F}{\partial x} & \frac{\partial F}{\partial y} \\ \frac{\partial G}{\partial x} & \frac{\partial G}{\partial y} \end{bmatrix}_{p^*} \begin{bmatrix} x - x^* \\ y - y^* \end{bmatrix}$$

The eigenvalues and eigenvectors of the Jacobian matrix, J, at an equilibrium point, p^*, determine the geometry of the phase portrait close to the equilibrium point p^*. These ideas can be extended to any autonomous system of ODEs. A parallel definition applies to a discrete dynamical system.

Linear pendulum Pendulum motion can be modeled by a nonlinear ODE, but there is an approximating linear ODE that works well for small angles θ, where $\sin \theta \approx \theta$. In that case, the mathematics is the same as that discussed for the mass on a spring in Module 4.

Linear system A linear system of first-order ODEs has only terms that are linear in the state variables. The coefficients can be constants or functions (even nonlinear) of t.

Example: Here is a linear system with state variables x and y, and constant coefficients $a, b, \ldots, h$:

$$x' = ax + by + c$$
$$y' = fx + gy + h$$

This can be written in matrix/vector form as:

$$z' = Az + k$$

$$z = \begin{bmatrix} x \\ y \end{bmatrix}, \qquad A = \begin{bmatrix} a & b \\ f & g \end{bmatrix}, \qquad k = \begin{bmatrix} c & h \end{bmatrix}$$

The example can be extended to n state variables and an $n \times n$ matrix A. If $z = p^*$ is an equilibrium point of a linear system, then $k = -Ap^*$ and the system may be written as

$$z' = A(z - p^*)$$

What is special about a constant coefficient linear system is that linear algebra can be applied to find the general solution. See General solution (for linear ODEs).

Lissajous figures Jules Antoine Lissajous was a 19th-century French physicist who devised ingenious ways to visualize wave motion that involves more than one frequency. For example, try plotting the parametric curve $x_1 = \sin 2t$, $x_2 = \sin 3t$ in the $x_1 x_2$-plane with $0 \leq t \leq 320$.

The graph of a solution $x_1 = x_1(t)$, $x_2 = x_2(t)$ of

$$\begin{bmatrix} x_1 \\ x_2 \end{bmatrix}'' = B \begin{bmatrix} x_1 \\ x_2 \end{bmatrix}, \qquad \text{for } B \text{ a } 2 \times 2 \text{ constant matrix}$$

in the $x_1 x_2$-plane is a Lissajous figure if the vector $(x_1(0), x_2(0))$ is not an eigenvector of B.

See also "Normal modes and frequencies."

Local IVP One-step methods for approximating solutions to the IVP

$$y' = f(t, y), \qquad y(t_0) = y_0$$

generate the $(n + 1)$st approximation, y_{n+1}, from the nth, y_n, by solving the local IVP

$$u' = f(t, u), \qquad u(t_n) = y_n$$

This is exactly the same ODE, but the initial condition is different at each step.

Logistic model The logistic equation is the fundamental model for population growth in an environment with limited resources. Many advanced models in ecology are based on the logistic equation.

For continuous models, the logistic ODE is

$$\frac{dP}{dt} = rP \left(1 - \frac{P}{K} \right)$$

where r is the intrinsic growth rate constant, and K is the carrying capacity.

For discrete models, the logistic map is

$$f_\lambda(x) = \lambda x \left(1 - \frac{x}{K} \right)$$

where λ is the intrinsic growth rate constant, and K is again the carrying capacity.

Mandelbrot Set In complex dynamics, for $f_c(z) = z^2 + c$, the Mandelbrot set is a bifurcation diagram in the complex c-plane, computed by coloring all c-values for which z does not iterate to infinity. It acts as a catalog of all the Julia sets for individual values of c.

The boundary of the Mandelbrot set is even more complicated than the boundary of a given Julia set. More detail appears at every level of zoom, but no two regions are exactly self-similar.

Two mathematicians at UCLA, R. Brooks and J. P. Matelsky, published the first picture in 1978. It is now called the Mandelbrot set, because Benoit Mandelbrot of the Thomas J. Watson IBM Research Center made it famous in the early 80s.

You can experiment with the Mandelbrot set in Module 13, on Screens 3.1 and 3.4.

Matrix An $n \times n$ square matrix A of constants, where n is a positive integer, is an array of numbers arranged into n rows and n columns. The entry where the ith row meets the jth column is denoted by a_{ij}.

In ODEs we most often see matrices A as the array of coefficients of a linear system. For example, here is a planar linear system with a 2×2 coefficient matrix A:

$$\begin{aligned} x' &= 2x - 3y \\ y' &= 7x + 4y \end{aligned} \qquad A = \begin{bmatrix} 2 & -3 \\ 7 & 4 \end{bmatrix}$$

Mixing A function $f : R \to R$ is "mixing" if given any two intervals I and J there exists an $n > 0$ such that the nth iterate of I intersects J.

Modeling A mathematical model is a collection of variables and equations representing some aspect of a physical system. In our case, the equations are differential equations. Steps involved in the modeling process are:

1. State the problem.
2. Identify the quantities to which variables are to be assigned; choose units.
3. State laws which govern the relationships and behaviors of the variables.
4. Translate the laws and other data into mathematical notation.
5. Solve the resulting equations.
6. Apply the mathematical solution to the physical system.
7. Test to see whether the solution is reasonable.
8. Revise the model and/or restate the problem, if necessary.

Moment of inertia The moment of inertia, I, of a body B about an axis is given by

$$I = \int \int \int_B r^2 \rho(x, y, z) \, dV(x, y, z)$$

where r is the distance from a general point in the body to the axis and ρ is the density function for B. Each

moment of inertia plays the same role as mass does in nonrotational motion, but, now, the shape of the body and the position of the axis play a role.

Newtonian damping A body moving through air (or some other medium) is slowed down by a resistive force that acts opposite to the body's velocity, v. In Newtonian damping (or Newtonian drag), the magnitude of the force is proportional to the square of the magnitude of the velocity, i.e., to the square of the speed:

$$\text{force} = -k|v|v \quad \text{for some positive constant } k$$

Newton's law of cooling The temperature, T, of a warm body immersed in a cooler outside medium of temperature T_{out} changes at a rate proportional to the temperature difference,

$$\frac{dT}{dt} = k(T_{out} - T)$$

where T_{out} is assumed to be unaffected by T (unless stated otherwise). The same ODE works if T_{out} is larger than T (Newton's law of warming).

Newton's second law Newton's second law states that, for a body of constant mass,

$$\text{mass} \cdot \text{acceleration} = \text{sum of forces acting on body}$$

This is a differential equation, because acceleration is the rate of change of velocity, and velocity is the rate of change of position.

Nodal equilibrium The behavior of the trajectories of an autonomous system of ODEs is nodal at an equilibrium point if all nearby trajectories approach the equilibrium point with definite tangents as $t \to +\infty$ (nodal sink), or as $t \to -\infty$ (nodal source).

If the system is linear with the matrix of coefficients A, then the equilibrium is a nodal sink if all eigenvalues of A are negative, and a nodal source if all eigenvalues are positive. This also holds at an equilibrium point of any nonlinear autonomous system, where A is the Jacobian matrix at the equilibrium point.

Nonautonomous ODE A system of ODEs with t occurring explicitly in the expressions for the rates is nonautonomous.

Nonlinear center point An equilibrium point of a nonlinear system, $x' = F(x, y)$, $y' = G(x, y)$, is a center if all nearby orbits are simple closed curves enclosing the equilibrium point.

Nonlinear ODE A nonlinear ODE or system has at least some dependent variables appearing in nonlinear terms (e.g., xy, $\sin x$, $\sqrt{x}$). Thus, linear algebra cannot be applied to the system overall. But, near an equilibrium (of which there are usually more than one for a nonlinear system of ODEs), a linearization is (usually) a good approximation, and allows analysis with the important roles of the eigenvalues and eigenvectors.

Nonlinear pendulum Newton's laws of motion give us

$$\text{force} = \text{mass} \times \text{acceleration}$$

In the circular motion of a pendulum of fixed length, L, at angle θ, acceleration is given by $L\theta''$. The only forces acting on the undamped pendulum are those due tension in the rod and gravity. The component of force in the direction in which the pendulum bob is moving:

$$F = mL\theta'' = -mg\sin\theta$$

where m is the mass of the pendulum bob, and g is the acceleration due to gravity. The mass of the rigid support rod is assumed to be negligible.

Normal modes and frequencies The normal modes of a second order system $z'' = Bz$ (where B is a 2×2 matrix with negative eigenvalues μ_1, μ_2) are eigenvectors v_1, v_2 of B. The general solution is all linear combinations of the periodic oscillations z_1, z_2, z_3, z_4 along the normal modes.

$$z_1 = v_1 \cos\omega_1 t, \quad z_2 = v_1 \sin\omega_1 t,$$
$$z_3 = v_2 \cos\omega_2 t, \quad z_4 = v_2 \sin\omega_2 t$$

where $\omega_1 = \sqrt{-\mu_1}$, $\omega_2 = \sqrt{-\mu_2}$, are the normal frequencies.

See also "Second order systems."

Normalized ODE In a normalized differential equation, the the highest order derivative appears alone in a separate term and has a coefficient equal to one.

ODE See ordinary differential equation.

On-off function See square wave.

Orbit See trajectory.

Order of the method A method of numerical approximation to a solution of an IVP is order p, if there exists a constant C such that

$$\max(|\text{global error}|) < Ch^p$$

as $h \to 0$.

Ordinary differential equation An ordinary differential equation (ODE) is an equation involving an unknown function and one or more of its derivatives. The order

of the ODE is the order of the highest derivative in the ODE. Examples:

$$\frac{dy}{dt} = 2t, \quad \text{(first-order, unknown } y(t)\text{)}$$

$$\frac{dy}{dt} = 2y + t, \quad \text{(first-order, unknown } y(t)\text{)}$$

$$x'' - 4x' + 7x = 4\sin 2t, \quad \text{(second-order, unknown } x(t)\text{)}$$

Oscillation times Oscillation times of a solution curve $x(t)$ of an ODE that oscillates around $x = 0$ are the times between successive crossings of $x = 0$ in the same direction. If the solution is periodic, the oscillation times all equal the period.

Oscillations A scalar function $x(t)$ oscillates if $x(t)$ alternately increases and decreases as time increases. The oscillation is periodic of period T if $x(t + T) = x(t)$ for all t and if T is the smallest positive number for which this is true.

Parametrization Each coordinate of a point in space may sometime be given in terms of other variable(s) or parameter(s). A single parameter suffices to describe a curve in space. Two parameters are required to describe a two-dimensional surface.

Period The period of a periodic function $u(t)$ is the smallest time interval after which the graph of u versus t repeats itself. It can be found by estimating the time interval between any two corresponding points, e.g., successive absolute maxima.

The period of a cycle in a discrete dynamical system is the minimal number of iterations after which the entire cycle repeats.

Periodic phase plane The periodic xx'-phase plane for the pendulum ODE $x'' + 0.1x' + \sin x = \cos t$ is plotted periodically in x. An orbit leaving the screen on the right comes back on the left. In other words, the horizontal axis represents $x \bmod 2\pi$. This view ignores how many times the pendulum bob has gone over the top. See Module 12, screen 1.4.

Phase angle The phase angle, δ, of the oscillatory function $u(t) = A\cos(\omega_0 t + \delta)$ shifts the the graph of $u(t)$ from the position of a standard cosine graph $u = \cos\omega_0 t$ by the amount δ/ω_0. The phase angle may have either sign and must lie in the interval $-\pi/\omega_0 < \delta < \pi/\omega_0$.

Phase plane The phase plane, or state plane, is the xy plane for the dependent variables x and y of the system

$$x' = F(x, y)$$
$$y' = G(x, y)$$

The trajectory, or orbit, of a solution

$$x = x(t), \qquad y = y(t)$$

of the system is drawn in this plane with t as a parameter. A graph of trajectories is called a phase portrait for the system.

The higher dimensional analog is called phase space, or state space.

Pitch The pitch (frequency) of an oscillating function $u(t)$ is the number of oscillations per unit of time t.

Poincaré Henri Poincaré (1854–1912) was one of the last mathematicians to have a universal grasp of all branches of the subject. He was also a great popular writer on mathematics. Poincaré's books sold over a million copies.

Poincaré section A Poincaré section of a second order ODE $x'' = f(x, x', t)$, where f has period T in t, is a strobe picture of the xx'-phase plane that plots only the points of an orbit that occur at intervals separated by a period of T time units, i.e., the sequence of points

$$P_0 = (x(0), x'(0))$$
$$P_1 = (x(T), x'(T))$$
$$\vdots$$
$$P_n = (x(nT), x'(nT))$$
$$\vdots$$

This view of phase space was developed by Henri Poincaré in the early twentieth century, because it is especially useful for analyzing nonautonomous differential equations. For further detail, see the entire second submodule of Module 12, Chaos and Control.

A Poincaré section is a two-dimensional discrete dynamical system. Another example of such a system is discussed in some detail in the second submodule of Module 13.

Population quadrant In a two-species population model, the population quadrant of the phase plane is the one where both dependent variables are non-negative.

Post-image In a discrete dynamical system, a post-image of a set S_0 is another set of points, S_1, where the iterates of S_0 land in one step.

For a Poincaré section of an ODE, S_1 would be the set of points arriving at S_1 when the ODE is solved from S_0 over one time period of the Poincaré section. See submodule 3 of Module 12.

Pre-image In a discrete dynamical system, a pre-image of a set S_0 is another set of points, S_{-1}, that iterate to S_0 in one step.

For a Poincaré section of an ODE, S_{-1} would be the set of points arriving at S_0 when the ODE is solved from S_{-1} over one time period of the Poincaré section. See submodule 3 of Module 12.

Products The products of a chemical reaction are the species produced by a reaction step. The end products are the species that remain after all of the reaction steps have ended.

Proportional Two variables are proportional if their ratio is constant. Thus, the circumference, c, of a circle is proportional to the diameter, because $c/d = \pi$.

The basic linear differential equation

$$\frac{dy}{dt} = ky$$

represents a quantity y whose derivative is proportional to its value.

Random Random motion is the opposite of deterministic motion. In random motion, there is no way to predict the future state of a system from knowledge of the initial state. For example, if you get heads on the first toss of a coin, you cannot predict the outcome of the fifth toss.

Rate constant Example: The constant coefficients a, b, and c in the rate equation

$$x'(t) = ax(t) - by(t) - cx^2(t)$$

are often called rate constants.

Rates of chemical reactions The rate of a reaction step is the speed at which a product species is created or (equivalently) at which a reactant species is destroyed in the step.

Reactant A chemical reactant produces other chemicals in a reaction.

Resonance This phenomenon occurs when the amplitude of a solution of a forced second order ODE becomes either unbounded (in an undamped ODE) or relatively large (in a damped ODE) after long enough times.

Rotation system As Lagrange discovered in the 18th century, the equations of motion governing a gyrating

body are

$$\omega_1' = \frac{(I_2 - I_3)\omega_2\omega_3}{I_1}$$

$$\omega_2' = \frac{(I_3 - I_1)\omega_1\omega_3}{I_2}$$

$$\omega_3' = \frac{(I_1 - I_2)\omega_1\omega_2}{I_3}$$

where I_j is the principal moment of inertia, and ω_j is the component of angular velocity about the jth body axis.

Saddle An equilibrium point of a planar autonomous ODE, or a fixed point of a discrete two-dimensional dynamical system, with the property that, in one direction (the unstable one), trajectories move away from it, while, in another direction (the stable one), trajectories move toward it.

At a saddle of an ODE, one eigenvalue of the associated linearization matrix must be real and positive, and at least one eigenvalue must be real and negative.

Scaling Before computing or plotting, variables are often scaled for convenience.

See also "Dimensionless variables."

Second order systems Second order systems of the form $z'' = Bz$ often arise in modeling mechanical structures with no damping, (and hence, no loss of energy). Here, z is an n-vector state variable, z'' denotes d^2z/dt^2, and B is an $n \times n$ matrix of real constants.

Although numerical solvers usually require that we introduce $v = z'$ and enter the system of $2n$ first order ODEs, $z' = v$, $v' = Bz$, we can learn a lot about solutions directly from the eigenvalues and eigenvectors of the matrix B.

See also "Normal modes and frequencies" and Screen 3.4 in Module].

Sensitivity An ODE model contains elements, such as initial data, environmental parameters, and functions, whose exact values are experimentally determined. The effect on the solution of the model ODEs when these factors are changed is called sensitivity.

Sensitivity to initial conditions A dynamical system has sensitive dependence on initial conditions if every pair of nearby points eventually gets mapped to points far apart.

Separatrix Separatrices are trajectories of a planar autonomous system that enter or leave an equilibrium point p with definite tangents as $t \to \pm\infty$, and divide

a neighborhood of p into distinct regions of quite different long-term trajectory behavior as t increases or decreases.

For more on separatrices see "Separatrices and Saddle Points" in Chapter 7.

Sink A sink is an equilibrium point of a system of ODEs, or a fixed point of a discrete dynamical system, with the property that all trajectories move toward the equilibrium.

If all eigenvalues of the associated linearization matrix at an equilibrium of a system of ODEs have negative real part, then the equilibrium is a sink.

Slope The slope of a line segment in the xy-plane is given by the formula

$$m = \frac{\text{change in } y}{\text{change in } x}$$

The slope of a function $y = f(x)$ at a point is the value of the derivative of the function at that point.

Slope field See direction field.

Solution A solution to a differential equation is any function which gives a true statement when plugged into the equation. Such a function is called a particular solution. Thus,

$$y = t^2 - 2$$

is a particular solution to the equation

$$\frac{dy}{dt} = 2t$$

The set of all possible solutions to a differential equation is called the general solution. Thus,

$$y = t^2 + C$$

is the general solution to the equation

$$\frac{dy}{dt} = 2t$$

Source A source is an equilibrium of a system of ODEs, or a fixed point of a discrete dynamical system, with the property that all trajectories move away from the equilibrium.

If all eigenvalues of the associated linearization matrix at an equilibrium of a system of ODEs have positive real part, then the equilibrium is a source.

Spiral equilibrium An equilibrium point of a planar autonomous system of ODEs is a spiral point if all nearby orbits spiral toward it (or away from it) as time increases.

If the system is linear with the matrix of coefficients A, then the equilibrium is a spiral sink if the eigenvalues of A are complex conjugates with negative real part, a spiral source if the real part is positive. This also holds at an equilibrium point of any nonlinear planar autonomous system, where A is the Jacobian matrix at the equilibrium.

Spring A Hooke's law restoring force (proportional to displacement, x, from equilibrium) and a viscous damping force (proportional to velocity, but oppositely directed) act on a body of mass m at the end of spring. By Newton's Second Law,

$$mx'' = -kx - bx'$$

where k and b are the constants of proportionality.

Spring force The spring force is often assumed to obey Hooke's law—the magnitude of the force in the spring is proportional to the magnitude of its displacement from equilibrium, and the force acts in the direction opposite to the displacement.

The proportionality constant, k, is called the spring constant. A large value of k corresponds to a stiff spring.

Square wave An on-off function (also called a square wave) is a periodic function which has a constant nonzero value for a fraction of each period; otherwise, it has a value of 0. For example, $y = A\text{SqWave}(t, 6, 2)$ is a square wave of amplitude A and period 6, which is "on" for the first 2 units of its period of 6 units, then is off the next 4 time units.

Stable An equilibrium point p^* of an autonomous system of ODEs is stable if trajectories that start near p^* stay near p^*, as time advances. The equilibrium point $p^* = 0$ of the linear system $z' = Az$, where A is a matrix of real constants, is stable if all eigenvalues of A are negative or have negative real parts.

State space The phase plane, or state plane, is the xy-plane for the dependent variables x and y of the system

$$x' = F(x, y)$$
$$y' = G(x, y)$$

The trajectory, or orbit, of a solution

$$x = x(t), \qquad y = y(t)$$

of the system is drawn in this plane with t as a parameter. A graph of trajectories is called a phase portrait for the system.

The higher dimensional analog is called phase space, or state space.

State variables These are dependent variables whose values at a given time can be used with the modeling ODEs to determine the state of the system at any other time.

Steady state A steady state of a system of ODEs is an equilibrium position where no state variable changes with time.

Surface A surface of a three-dimensional object is just its two- dimensional "skin," and does not include the space or volume enclosed by the surface.

Taylor remainder For an $n + 1$ times differentiable function $u(t)$, the difference (or Taylor remainder)

$$u(t) - [u(t_0) + hu'(t_0) + \cdots + \frac{1}{n!}h^n u^{(n)}(t_0) + \cdots]$$

can be written as

$$\frac{1}{(n+1)!}h^n u^{(n+1)}(c)$$

for some c in the interval $[t_0, t_0 + h]$, a fact which gives useful estimates.

Taylor series expansion For an infinitely differentiable function $u(t)$, the Taylor series expansion at t_0 for $u(t_0 + h)$ is

$$u(t_0) + hu'(t_0) + \frac{1}{2}h^2 u''(t_0) + \cdots + \frac{1}{n!}h^n u^{(n)}(t_0) + \cdots$$

Taylor series method Look at the IVP $y' = f(t, y)$, $y(t_0) = y_0$. For a step size h, the three-term Taylor series method approximates the solution $y(t)$ at $t_{n+1} = t_n + h$, for $n = 0, 1, 2, \ldots$, using the algorithm

$$y_{n+1} = y_n + hf(t_n, y_n) + \frac{1}{2}h^2 f_t(t_n, y_n)$$

Trace The trace of a square matrix is the sum of its diagonal entries. So

$$\text{tr}\begin{bmatrix} a & b \\ c & d \end{bmatrix} = a + d$$

Trace-determinant parabola The eigenvalues λ_1, λ_2 of a 2×2 matrix A are given by

$$\lambda_1, \lambda_2 = \frac{\text{tr}\, A \pm \sqrt{\text{tr}^2 A - 4 \det A}}{2}$$

The trace-determinant parabola, $4 \det A = \text{tr}^2 A$, divides the $\text{tr}\, A - \det A$ plane into the upper region where A's eigenvalues are complex conjugates and the lower region where they are real. The two eigenvalues are real and equal on the parabola.

Trajectory A trajectory (or orbit, or path) is the parametric curve drawn in the xy-plane, called the phase plane or state plane, by $x = x(t)$ and $y = y(t)$ as t changes, where $x(t)$, $y(t)$ is a solution of

$$x' = F(x, y, t)$$
$$y' = G(x, y, t)$$

The trajectory shows how $x(t)$ and $y(t)$ play off against each other as time changes.

For a higher dimensional system, the definition extends to parametric curves in higher dimensional phase space or state space.

Unstable An equilibrium point p^* of an autonomous system of ODEs is unstable if it is not stable. That means there is a neighborhood N of p^* with the property that, starting inside each neighborhood M of p^*, there is at least one trajectory that goes outside N as time advances.

Vector A vector is a directed quantity with length. In two dimensions, a vector can be written in terms of unit vectors $\hat{\mathbf{i}}$ and $\hat{\mathbf{j}}$, directed along the positive x and y axes.

Viscous damping A body moving through air (or some other medium) is slowed down by a resistive force that acts opposite to the body's velocity, v. In viscous damping (or viscous drag), the force is proportional to the velocity:

$$\text{force} = -kv$$

for some positive constant k.

Wada property The Wada property, as described and illustrated on Screen 3.2 of Module 12 is the fact that:

> Any point on the boundary of any one of the areas described on Screen 3.2 is also on the boundary of all the others.

The geometry/topology example constructed by Wada was the first to have this property; we can now show that the basins of attraction for our forced, damped pendulum ODE have the same property. See Module 12 and Chapter 12.

All we know about Wada is that a Japanese manuscript asserts that someone by that name is responsible for constructing this example, showing that for three areas in a plane, they can become so utterly tangled that every boundary point touches all three areas!